Greening the Globe

Recent decades have seen a rapid expansion of environmental activity in the world, including the signing of a growing number of environmental treaties and the formation of international organizations like the United Nations Environment Programme (UNEP). *Greening the Globe* employs world society theory (also known as world polity theory or sociological institutionalism) to explore the origins and consequences of international efforts to address environmental problems. Existing scholarship seems paradoxical: case studies frequently criticize treaties and regulatory structures as weak and ineffective, yet statistical studies find improvements in environmental conditions. This book addresses this paradox by articulating a Bee Swarm model of social change. International institutions rarely command the power or resources to impose social change directly. Nevertheless, they have recourse via indirect mechanisms: setting agendas, creating workspaces where problems can be addressed, empowering various pro-environmental agents, and propagating new cultural meanings and norms. As a result, world society generates social change even if formal institutional mechanisms and sanctions are weak.

Ann Hironaka is an associate professor of sociology at the University of California, Irvine. She studies environmental sociology, politics, and war from a global perspective. Her research examines the historical emergence of the global environmental regime and its impact on national policy and environmental practices around the world. Her work on environmentalism has appeared in the *American Sociological Review, International Organization,* and *Social Forces.* She is also a member of the American Sociological Association Task Force on Climate Change. Hironaka's first book, *Neverending Wars* (2005), examines the intractable civil wars of the contemporary era and the role of the international community in perpetuating these conflicts.

Greening the Globe

World Society and Environmental Change

ANN HIRONAKA

University of California, Irvine

CAMBRIDGE
UNIVERSITY PRESS

32 Avenue of the Americas, New York NY 10013-2473, USA

Cambridge University Press is part of the University of Cambridge.

It furthers the University's mission by disseminating knowledge in the pursuit of education, learning and research at the highest international levels of excellence.

www.cambridge.org
Information on this title: www.cambridge.org/9781316608425

First published 2014
First paperback edition 2016

A catalogue record for this publication is available from the British Library

Library of Congress Cataloguing in Publication data
Hironaka, Ann.
Greening the globe : world society and environmental change / Ann Hironaka.
pages   cm
Includes bibliographical references and index.
ISBN 978-1-107-03154-8 (hardback)
1. Environmentalism.   2. Social change.   3. Environmental protection – International cooperation.   4. Environmental policy – International cooperation.   I. Title.
GE195.H566   2014
363.7–dc23   2014008510

ISBN 978-1-107-03154-8 Hardback
ISBN 978-1-316-60842-5 Paperback

To John Meyer, with thanks

Contents

Figures and Tables

Figures

Tables

Preface and Acknowledgments

I first encountered John Meyer when I was a Stanford undergraduate student as I was embarking upon my first course in sociology. When I began to look at graduate schools, I was surprised to learn that John Meyer's perspective was not common within sociology. Having encountered John in my first venture, I had naively supposed that he broadly represented the discipline of sociology. Thus I returned to Stanford for graduate work in sociology, figuring I could easily spend years trying to figure out what John was saying. I am grateful for the first-rate methodological and theoretical training that John imparts to his students in an offhand way. I also feel privileged to have experienced firsthand his incomprehensible scribbles on the chalkboard (in the days before he began providing scribbles in PowerPoint), and the way in which he gazes out the window before delivering a particularly sharp or amusing insight.

Moreover, I am grateful to John for encouraging me to extend world society theory into the darker realm of violence and war. At times, I have felt like a black sheep for venturing beyond the happy world society domain of human rights and educational progress. Yet I now suspect that John has enfolded many black sheep into his flock, thanks to his intellectual breadth and generosity.

I am also grateful to be embraced by the equally generous community of Meyer students and collaborators. In particular, I thank John Boli, David Frank, John Meyer, Evan Schofer, David Strang, and George Thomas, who read and commented extensively on the manuscript, as well as Elizabeth Boyle, Gili Drori, Wesley Longhofer, David Meyer, Nolan Philips, Francisco Ramirez, Amber Tierney, and members of the Irvine Comparative Sociology Workshop and the Social Movements and Social

Justice Workshop at UC Irvine. Special thanks are directed to Wesley Longhofer for designing the figures and to Evan Schofer for his extensive editing efforts and his help with the statistical analyses. Thanks also to the many friends and colleagues who contributed intellectually to this project: Pertti Alasuutari, Kenneth "Andy" Andrews, Colin Beck, Jeff Broadbent, Wade Cole, Riley Dunlap, Erin Evans, Dana Fisher, Marion Fourcade, Joe Galaskiewicz, Selina Gallo-Cruz, Michael Goldman, Ralph Hosoki, Ron Jepperson, Andrew Jorgenson, Mike Landis, Karen Robinson, Joachim Savelsberg, Kristen Shorette, Kiyo Tsutsui, Marc Ventresca, and Christine Min Wotipka.

I

World Society and Social Change

Efforts to address environmental problems have grown tremendously over the past several decades. Environmental concerns now encompass the breadth of human relationships with nature – from air and water pollution, to habitat loss and biodiversity, to climate change. The international community has established a broad array of environmental treaties covering many domains. Countries around the world have enacted environmental laws and created environmental ministries. Innumerable citizen groups have mobilized to address local, national, and planetary environmental issues.

It is easy to take these social changes for granted, as a necessary and obvious response to the environmental disasters around us. Yet the growth of pro-environmental concern and activity was by no means foreordained. A basic sociological insight is that social problems, however egregious, do not necessarily prompt effective solutions. The oppressed too rarely rise up against their oppressors. Likewise, environmental destruction can occur on a massive scale without a social response (Ponting 1991). Scholars have pointed out many obstacles to change, including powerful interests that may benefit from environmentally destructive practices and the complexities of developing effective environmental governance (Espeland 1998; Young 1989b).

World society theory[1] is a perspective that can be used to make sense of the massive expansion of environmentalism around the globe. World

---

[1] World society theory is known by a variety of labels, including neo-institutional theory, world polity theory, sociological institutionalism or the Stanford school of institutionalism. This branch of theory is seen as distinct from economic institutionalism (North 1990), political institutionalism (Amenta 2006; Amenta and Caren 2004; Jenkins and

society theory argues that the behavior of nations, organizations, and individuals can be understood as deriving from a common global culture, often formally codified and institutionalized within international treaties and organizations. Cultural understandings are fundamental to social change, once they become widely accepted across the international community and incorporated into global and national governance. Governments around the world ultimately address environmental concerns as a result of growing attention to environmental issues in the international sphere and shared cultural understandings that have rendered environmental issues salient. This macro-cultural perspective provides an alternative to explanations of environmentalism that focus on the growing urgency of environmental threats, changes in the attitudes of individuals, or the dynamics of local or national politics.

Empirical research in the world society tradition observes that the emergence of pro-environmental organizations, treaties, and cultural meanings in the international sphere has encouraged governments to adopt a wide range of pro-environmental policies and laws (Hironaka and Schofer 2002; Frank, Hironaka, and Schofer 2000a; Frank et al. 1999; J. Meyer et al. 1997b). However, scholars have questioned the impact of international pro-environmental policies and laws in actually improving environmental conditions on the ground (Buttel 2000). For instance, despite the fanfare surrounding the 1992 United Nations Framework Convention on Climate Change, environmental damage continues at a rapid pace (Paterson 1996; Falkner 2008). In the language of organizational sociology, regulatory structures appear merely ceremonial and only loosely coupled from intended outcomes (J. Meyer and Rowan 1977).

In response, scholars in the world society tradition have increasingly turned their attention to empirical outcomes and practices in addition to policies (Cole and Ramirez 2013; Shorette 2012; Schofer and Hironaka 2005; Hafner-Burton and Tsutsui 2005). This emerging literature has several notable findings that set the stage for this book. First, a number of studies observe powerful effects of world society on actual social practices around the globe – including studies specifically looking at the global environmental regime (Shorette 2012; Hadler and Haller 2011;

Form 2005), and historical institutionalism (Skocpol 1992; Hicks 1999). See Amenta and Ramsey (2010) for more detailed discussion. For recent reviews see J. Meyer 2010; Schofer et al. 2012. World society theory is also distinct from world-systems theory, which is a global perspective that focuses principally on economic processes (Wallerstein 1999; Chirot 1986).

Shandra, Leckband, and London 2009a; Shandra et al. 2009b; Shandra, Shor, and London 2008; Jorgenson 2007, 2006, 2003; Schofer and Hironaka 2005). States that are embedded in world society tend to show improvement in environmental conditions such as deforestation, greenhouse gas production, chlorofluorocarbon production, and pesticide use. International treaties and regimes tend to be more consequential as they become increasingly elaborated over time (Cole and Ramirez 2013; Shorette 2012; Schofer and Hironaka 2005). However, these processes do not always go smoothly, and improvements are rarely as substantial as hoped (Hafner-Burton and Tsutsui 2005; Shorette 2012). Global efforts have slowed the accelerating pace of environmental damage in many cases, but environmental problems are far from solved. In general, however, empirical studies have shown that global regimes can have an ameliorating impact on outcomes on the ground. These findings are briefly reprised in the Appendix.

This book explores how and when international institutions generate effective social change, with a focus on improvements in environmental conditions. Drawing on cases and historical examples, the world society account of social change is unpacked. World society scholarship is best known for quantitative studies that focus on large-scale correlations. This book spells out the processes and mechanisms underlying these correlations to explain how world society can simultaneously give rise to loose coupling and also yield long-term social change.

To foreshadow the argument, international institutions – and the cultural meanings they relay – powerfully enable social change but not necessarily in a direct or tightly coupled way. Institutions define new social problems and cultural meanings, create workspaces in which actors can address those problems, and empower individual or organizational "agents" who work toward addressing environmental issues. This perspective counters the conventional view that environmentalism is a natural and inevitable reaction to severe environmental problems within Western countries in the late 1960s (Brenton 1994; Rowland 1973). At the same time, world society theory is a counterpoint to materialist theories that see no possibility for social change in a capitalist world economy.

World society theory is a form of institutional analysis that focuses on both cultural and organizational processes. International regimes coevolve with foundational cultural shifts in meaning. Accepted social understandings and cultural frames regarding the environment have changed radically over the past century, with profound implications for society. The environment is increasingly characterized as interrelated interactions of planetary scope that humans can disrupt or protect, and this view has become embedded in social discourses and practices. At the same time, pro-environmental meanings have proliferated in a broad array of national and local contexts. These cultural changes are neither inevitable nor unopposed. Alternative institutional structures and cultural frames provide the basis for resistance and opposition. Nevertheless, rapidly expanding global activity around environmental issues has proven surprisingly consequential, and resistance in many cases has been swamped by the broader sea change in cultural meanings and institutional activity.

This book develops the world society approach to environment protection, fleshing out the argument and mechanisms and extending the perspective in several ways. The book takes on four major issues, each explored through a brief case study. First, the book examines the origin of the global environmental regime. World society theory has most often been employed to explain patterns of global diffusion. The question of institutional origins – where international institutions come from – has been given less attention (Schneiberg and Clemens 2006, but see Cole 2011). Chapter 2 addresses the emergence of the environmental regime, focusing on the events surrounding the United Nations Conference on the Human Environment at Stockholm in 1972 and the formation of the United Nations Environment Programme (UNEP).

Second, the book explores the effects of institutional structures on concrete outcomes – actual changes in environmental pollution and degradation. World society theory is very good at explaining policy diffusion, with the caveat that policies are sometimes disconnected from actual social change on the ground. Chapter 3 considers when and how institutional structures prompt substantive change, focusing on the global regime on ozone-depleting substances in the 1970s and 1980s.

Third, the book explores the concept of "agents," an alternative to the more conventional term "actor." World society theory is an aggressively structural approach that de-emphasizes conventional notions of social actors and agency. Chapter 4 explores the role of organizational agents for environmental protection, including international nongovernmental organizations (INGOs) and social movements, with an examination of the case of hazardous waste in the United States and elsewhere from the 1970s to the 1990s.

Finally, the book explores the role of cultural meaning and evolving patterns of conflict in world society theory. World society theory, more than other institutional perspectives, stresses the importance of culture, ideas, and meaning. Chapter 5 examines evolving cultural meanings, and their implications for contestation and conflict, focusing on efforts to create a global regime to address climate change from 1992 to the present.

The rest of this introductory chapter articulates a world society model of social change that explains how the seemingly weak influences of international institutions and global culture can nevertheless generate dramatic changes in activity around the globe. This approach focuses on myriad loosely coupled factors rather than treating behavior as a direct and proximate consequence of specific treaties or policies and their implementation. Next, the chapter contrasts the world society perspective with dominant explanations of social change in the literature, including modernization theories and social movement arguments. The chapter concludes by outlining the overall structure of the book.

World Society and Social Change: The Strength of Weak Mechanisms

A conundrum for scholars seeking to explain major historical transformations is that social change appears to be halting, contested, and partial when closely scrutinized. Consider the topics of democratization and gender equality. Close analyses of any democratic society are likely to find failures of effective democracy beneath the facade. Likewise, ongoing gender discrimination can be uncovered even in the most egalitarian societies. Yet when one looks at the forest rather than the trees, tremendous social change has occurred. The past century has seen multiple waves of democratization and huge increases in the social status of women across much of the globe (C. Beck 2011; Ramirez, Soysal, and Shanahan 1997). World society theory provides imageries and arguments that explain how seemingly fragmentary and halting efforts to address environmental concerns may nevertheless reflect a broader sea change that is dramatic when placed in historical context. The concept of "loose coupling," adopted from organizational theory, helps characterize the uneven patterns of emergent global change. The loose coupling between policies and outcomes does not imply that change never occurs. Instead, loose coupling suggests that change is not the result of tightly coordinated organizational actions.

Much research implicitly adopts what might be dubbed the Smoking Gun model of social change. Scholars strive to identify tight sequences

Policy ————————→ Intended outcome

FIGURE 1.1. The Smoking Gun: A tightly coupled model of social change.

of cause and effect, linking specific environmental policies or pro-environmental groups to improved environmental conditions. Such studies turn their analytical lens toward a specific path of social change, such as the link between treaty ratification and environmental improvements on the ground. In order to formulate a tractable research question, complex combinatorial dynamics – for instance, where myriad weak factors combine to affect a given outcome – are set aside in favor of relatively parsimonious models. Figure 1.1 outlines the causal reasoning of the Smoking Gun model.

Following the Smoking Gun model, environmental scholars have sought to trace the influence of particular policies or organizations on their intended environmental outcomes. In many cases, the direct effect of policy or organization turns out to be negligible. For instance, international treaties and national legislation often fail to promote improved environmental practices (Susskind 1994; Keohane and Levy 1996; Hurrell and Kingsbury 1992; Glennon and Stewart 1998). Governmental environmental agencies and ministries are frequently ineffective at fulfilling their legislative mandates (Mazmanian and Nienaber 1979; Mazmanian and Sabatier 1983; R. Andrews 2006). And only on rare occasions do social movement groups influence environmental outcomes (Szasz 1994; Giugni 2004). Thus many studies dourly conclude that many environmental laws, treaties, and policies are failures. This empirical literature is discussed further in Chapter 3.

Even when overall improvements in environmental conditions are observed, it can be quite difficult to demonstrate the importance of a specific treaty, policy, organization, or social movement. In the case of deforestation, for instance, studies have demonstrated that organizations and social movements affected forestry policies in a single locale, but cannot conclusively establish systematic effects on deforestation at the national level (Gellert 2010). Given the magnitude of planetary environmental problems, such local efforts appear insignificant. As one scholar morosely observed, "Nobody can know for sure that their advice was the turning point over some issue, so the impression of having an influence is never confirmed" (Barratt 1996, 9).

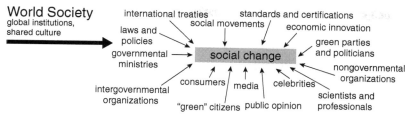

FIGURE I.2. The Bee Swarm: A loosely coupled model of social change.

This book outlines an alternative model of social change resulting from numerous factors operating at multiple levels of analysis – none of which may be necessary or sufficient. This approach draws on the concept of loose coupling from the organizational literature. In a loosely coupled world, widespread social change may result from the aggregation of many disparate mechanisms. These weak factors might be thought of as akin to a swarm of bees. A single bee sting may be a minor irritant, but the effects of a bee swarm can be deadly. In a Bee Swarm model of social change, a dense network of causal factors shapes outcomes, even if most individual causal factors prove weak or inconsequential. By the same token, the weakness of any particular causal factor does not necessarily mean that environmental conditions will fail to improve. In other words, social change is due to the aggregation of causal factors pushing in the same direction.[2] Figure I.2 outlines the weak causal forces of the Bee Swarm model.

This Bee Swarm model is inspired by the organizational concept of loose coupling (Bromley and Powell 2012; March 1981; J. Meyer and Rowan 1977; Weick 1976). Loose coupling should not be interpreted as a total disconnection between policy and outcome. Instead, the concept of loose coupling broadly refers to weak connections or control across an organization or social system. Loosely coupled systems may evolve in a disorganized fashion but can also undergo systematic change, for instance when a common external influence comes to bear on many parts of the system (J. Meyer and Rowan 1977; Schofer and Hironaka 2005).

Loose coupling occurs when organizations face a heterogeneous environment composed of multiple or overlapping regulatory structures, organizations, interests, and actors that are inconsistent or even contradictory.

---

[2] Thanks are directed to John Boli for suggesting the Bee Swarm imagery.

Various subunits of the organization respond to the hodgepodge of external demands in disparate and conflicting ways. For example, a corporation might face pressure to comply with environmental regulations yet may also face contradictory pressures to please stockholders, maintain a trendy image with the public and decrease workforce turnover. Under such conditions, the enactment of environmental regulations may have diverse effects throughout a single organization.

However, if pro-environmental cultural meanings become institutionalized in society, organizations may face increasingly consistent demands for environmental protection, from a variety of audiences. The hodgepodge of external pressures may coalesce into a more uniform chorus pushing in a common direction. Eventually the aggregated pressures may lead to improvement in environmental practices even if no single factor is decisive. Under these circumstances, change may occur despite weak internal linkages in a social system. Sometimes practices change before policies; other times change follows policy reform. Even in the latter case, one should be wary of assuming a tight linkage, as changed practices may be the product of broad external pressures rather than a response to formal policies or rules. This type of loosely coupled systemic shift has been termed *institutional drift* (J. Meyer and Rowan 1977; Schofer and Hironaka 2005; see also Mahoney and Thelen 2010). The power of the institutional context is not in a particular formal policy but in the cumulative effects of multiple indirect and loosely coordinated influences. As environmentalism becomes deeply institutionalized in the culture of global and national societies, pro-environmental activity becomes routine.

Culture is central to world society arguments, and often plays a critical role in animating and directing Bee Swarms in a manner that generates widespread social change. Cultural meanings orient members of society toward common issues or social problems. In the case of the environment, the cultural meaningfulness of environmental protection is no longer limited to scientific experts or particular issues, but has broadly diffused across the globe (Dunlap and Mertig 1992). People in all walks of life now perceive the meaningfulness and relevance of environmental issues. The sustainability of environmental practices in Brazil or New Zealand can call up fervent debates in North American grocery stores. This broad expansion of the cultural meaning of the environment provides a uniform orientation for the Bee Swarm, increasing the impact of each weak influence.

The Bee Swarm imagery is oversimplified in that it implies that the effects of bees are simply cumulative. In fact, they often reinforce or amplify one another. Pro-environmental laws, for instance, may prove more consequential when accompanied by environmental advocacy groups, and vice versa. And both laws and environmental groups are likely to have more traction in historical periods when pro-environmental meanings and principles are institutionalized in international treaties and organizations. The broad orientation of the Bee Swarm has greater impact than the efforts of any single bee.

These two differing models of social change motivate different empirical approaches. Scholarship drawing on the Smoking Gun imagery of change focuses on identifying proximate factors, such as a specific treaty, law, or social movement, that have strong causal impacts on a particular environmental outcome. In contrast, world society theory shifts attention toward the broad social context – the changing cultural and institutional environment that animates the Bee Swarm. Any single law, organization, or pro-environmental organization may prove inconsequential for predicting a given outcome. Moreover, it is usually impossible to enumerate and measure all of the possible mechanisms. The historical emergence and institutionalization of the global environmental regime in world society is the starting point for understanding the growing Bee Swarm of mechanisms ultimately leading toward social change.

Explaining Global Environmentalism

As recently as fifty years ago, most nations had no significant rules regarding what substances might be dumped in the ocean. Few attempts were made to control smoke and pollutants that were put into the air, despite the infamous fogs and smogs in London and Los Angeles (Brimblecombe 1987; Hironaka 2003). In the United States, federal laws regulating basic issues of air pollution, water pollution, and waste disposal were minimal or absent until around 1970 (R. Andrews 2006). By contrast, we now live in a world in which environmental regulation has become routine. These regulations may still be far from the ideal proposed by environmentalists, but it is a different world compared to the days when cities and corporations thought nothing of dumping untreated waste into waterways and people casually tossed trash out the windows of their cars.

What caused this shift toward greater environmental protection? The following section discusses three theoretical traditions in the scholarly literature: (1) modernization theory, (2) social movements, as a potential counterbalance to capitalist interests, and (3) world society arguments.

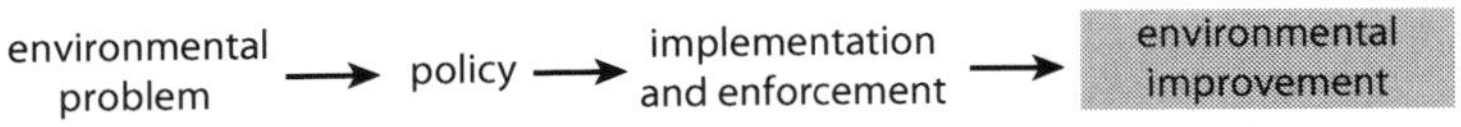

Greening the Globe

environmental problem $\longrightarrow$ policy $\longrightarrow$ implementation and enforcement $\longrightarrow$ environmental improvement

FIGURE 1.3. The modernization model of environmental change.

Modernization Theory

Ecological modernization theory, which derives from classical modernization theory, offers a hopeful viewpoint regarding environmental change (York, Rosa, and Dietz 2003; Mol 2001; U. Beck 1999; Inglehart 1977; Weidner and Jänicke 2002; Fisher and Freudenburg 2002; Fisher and Freudenburg 2001; Jänicke and Weidner 1996; Jänicke and Weidner 1995; Sonnenfeld 2000; Spaargaren and Mol 1992). Ecological modernization theorists argue that advanced capitalism provides the resources and technologies to combat environmental problems while the sophisticated institutional infrastructures of capitalist democracies support effective policies and regulation that reduce environmental destruction. This work dovetails with the post-materialist tradition (Inglehart 1977), which attends to value changes associated with economic modernization. Specifically, affluence allows people to transcend immediate economic concerns and shift toward "post-materialist" values that include greater support protecting the environment. Figure 1.3 provides a causal model of the processes outlined by ecological modernization theory.

In ecological modernization theory, social change is a response to environmental threats that are either apparent to all or are identified by scientists and experts. Affluence is an important mediating factor. Poor societies lack the resources to effectively address environmental issues, and individuals are more concerned with economic survival than with pro-environmental initiatives. Once sufficient levels of wealth are achieved, however, societal values shift in favor of environmental protection. Citizens vote for pro-environmental politicians and are willing to alter consumption habits. In the end, pro-environmental shifts in social values translate into improved environmental protection.

Science plays a major role in the identification of objective environmental problems and the development of solutions. As scientific knowledge accrues, an "epistemic community" is developed in which scientific experts build improved understandings of environmental issues (P. Haas 1992; E. Haas, Williams, and Babai 1977; Litfin 1994; Caldwell 1990). For instance, scientists initially discovered the deleterious effects of chemicals such as DDT and chlorofluorocarbons (CFCs) and theorized the potentially catastrophic effects of the buildup of greenhouse gas emissions. As scientists came to recognize the dangers these environmental conditions posed for humans, affluent societies became motivated to create solutions to these problems.

Modernization theories envision a world in which affluent societies gravitate toward rational solutions for the environmental problems uncovered by physical science. Treaties, legislation, and policies play a key role in mandating these solutions. If aerosol sprays are damaging the ozone layer, laws or treaties are put into place to eliminate the production of CFCs. Reasonable negotiation may ensure new legislation is not prohibitively costly for industrial groups or consumers. Technological innovation facilitates the process, as environmentally cleaner production processes reduce environmental damage and may also increase economic efficiency. In the long run, rational planning leads to appropriate regulation, ultimately resolving major environmental problems. As Spaargaren and Mol (1992, 334) argue, "ecological modernization indicates the possibility of overcoming the environmental crisis without leaving the path of modernization."

World society theory is sometimes lumped together with ecological modernization theory because both perspectives predict the spread of pro-environmental regulation. Despite superficial similarities, the underlying theories and mechanisms are fundamentally different. These differences are worked out in detail in Chapters 2 through 5, yielding an account of global environmentalism that departs from ecological modernization theory. In a nutshell, world society theory views environmental problems as socially constructed. From a world society perspective, the spread of ideas and culture is decisive, instead of industrialization, affluence, or material well-being.

The world society emphasis on loose coupling and diffuse mechanisms of cultural change provides an alternate explanation for policies that appear as failures from the modernization perspective. A great deal of research shows that environmental policies usually fail to produce the intended environmental outcomes (see Chapter 3). The modernization perspective acknowledges obstacles such as lack of resources, governmental capability, or political will. On a global scale, impoverished, corrupt, or war-torn governments may fail to address environmental problems, especially when compared to affluent countries (Jänicke 1996). But modernization theory has difficulty explaining how social change can occur if laws are routinely breached or when environmental reform is haphazard and poorly organized. Ecological modernization arguments are also unable to account for the rapid growth of environmental policy and pro-environmental attitudes in developing countries that are too impov-erished to afford the luxury of environmental health (Dunlap and Mertig 1992; Dunlap 1995). Unfortunately, these conditions of poor compliance and weak governmental capability tend to be the norm rather than the exception where environmental protection is concerned.

Capitalist Interests and Social Movement Responses

Classic scholarship on the environment, developed from Marxist theory, focuses on the role that capitalism and economic elites play in perpetuating the destruction of the natural environment. Marxist scholars have argued that capitalist economies require ever-increasing levels of consumption that inevitably conflict with the well-being of the natural environment (Schnaiberg 1980; O'Connor 1991; Gould, Pellow, and Schnaiberg 2008). From this perspective, substantive pro-environmental change is unlikely without a radical reorganization of modern industrial economies. World-system scholars have extended these ideas to address global capitalism, and in particular to explain how capitalist production may shift envi-ronmental destruction to the poorest of nations (Roberts and Grimes 2002; Bunker 1985; Jorgenson 2006; Jorgenson 2003; Shandra, Shor, and London 2009). Drawing on these themes, social movement research has sought to examine whether grassroots social movement groups may be capable of resisting the entrenched interests of capitalist elites (Pellow 2007; Giugni 2004).

Allen Schnaiberg outlined the classic argument that capitalist econo-mies are driven by the "treadmill of production" whereby demands for investment opportunities force increased production that requires ever increasing rates of consumption (Schnaiberg 1980; Schnaiberg and Gould 1994; Gould et al. 2008; Schnaiberg, Pellow, and Weinberg 2002). The ills of this system are easily apparent, such as worker exploitation and escalating inequality, as well as environmental degradation (O'Connor 1991; Pellow 2007; Pepper 1984). Scientists may identify new environ-mental problems and provide expert knowledge about more observable phenomena. Yet esoteric knowledge is not needed to recognize that the capitalist mode of consumption is unsustainable. Instead, the problem is that social change is opposed by powerful economic elites.

Building on this tradition, world-systems scholars have also shown that the Western industrial economies are the main perpetrators of global envi-ronmental degradation (Sklair 1994; Roberts and Grimes 2002; Bunker 1985; Chase-Dunn 1989; Grant, Bergesen, and Jones 2002; Wallerstein 1999; Bergesen and Parisi 1999; Bunker and Ciccantell 1999; Jorgenson

2006; Rudel 1993). Since the highly advanced capitalist economies often lack adequate domestic resources to fuel their own consumption, these resources must be provided by other parts of the world. Western capitalism uses its economic and political leverage to create structurally dependent states in the global South that are forced to supply resources at low cost. The rampant exploitation of these resources to supply the endless consumption of the West leads to environmental ills in the global South such as deforestation, displaced people, strip mining, and other significant environmental troubles (Bunker 1985; Bunker and Ciccantell 1999; Rudel 1993).

Marxist and world systems theories are generally pessimistic about the likelihood of improvement in environmental outcomes. Short of revolution, Marxist-based theories expect little actual improvement in environmental outcomes. Instead, ecological collapse seems the likely outcome. For instance, Gould et al. pessimistically predict, "If the treadmill *is* truly unsustainable both socially and ecologically, at some point it must either exhaust the planet's capacity to provide economically necessary resource pools and waste sinks or produce such deep, widespread social suffering that the vast majority forcibly dismantles it." (2008, 99).

Marxist-based theories have difficulty explaining the rapid growth of pro-environmental policy and activity since the 1970s. Some scholars argue that contemporary environmental efforts are superficial "window dressing" that leave the status quo largely unchanged. Frequent examples of environmental policy failure, noted above, can be seen as consistent with this view. Others argue that the improvements in environmental conditions principally benefit capitalist elites or the core nations of the world. Studies showing widespread change in pro-environmental attitudes and successful efforts at environmental change are harder to explain from this perspective (e.g., Schofer and Hironaka 2005; Shorette 2012).

Other researchers have turned to social movements as one potential mechanism to explain pro-environmental social change, in the absence of a full-blown revolution against capitalism. Drawing on a variety of tactics, environmental social-movement groups seek to expose the self-interest of economic elites and mobilize the quiet majority of supporters of environmental protection. However, there is little empirical research examining the effectiveness of social movements in motivating social change in terms of actual environmental conditions. Scholars have mainly examined the effect of social movements on various intermediate outcomes, such as increased access to the system, the addition of issues to the agenda, or the adoption of desired legislation (Burstein 1999; Giugni 2004; Amenta and

FIGURE 1.4. The social movement model of environmental change.

Caren 2004; Jenkins and Form 2005). Figure 1.4 outlines the processes suggested by social movement arguments.

This book focuses on social movement arguments as providing the most explicit model of social change from an interest-based perspective. Social movement scholars acknowledge that change is not easy in a world of powerful capitalist interests. Economic elites enjoy advantages in structural position and resources to block environmental protection efforts on a variety of fronts. Capitalist elites may adroitly prevent environmental issues from being placed on the political agenda or discussed in public forums (Lukes 1974; Crenson 1971). If legislation is proposed, corporate lobbyists can utilize their resources to influence policymakers to dilute or defeat the proposed bill. And if policies with teeth are enacted, corporations may simply ignore the law or petition for an exemption (Schnajberg et al. 2002). At every step, economic elites possess advantages that allow them to block or dilute pro-environmental efforts.

In the face of these obstacles, extraordinarily committed citizens or social movement organizations may emerge to advocate for environmental policies and raise public awareness to pass pro-environmental legislation. If environmental legislation is passed, social movement watchdogs may be needed to catch corporations that are evading regulations. Yet these citizen groups must fight an uphill battle at every step, lacking the structural advantages enjoyed by capitalist elites. Consequently, they must utilize tactics of the weak, such as protesting, organizing, or educating, in order to induce social change. As Gould et al. summarize, concerned citizens must "face the reality that these social forces were (and remain) at a major power disadvantage vis-à-vis political and economic elites" (2008, 16).

Although the social movement perspective suggests possible mechanisms for the improvement of environmental outcomes, the overall outlook is not rosy. While environmental groups may achieve minor victories, the power and influence of social movement organizations pales in comparison to the overwhelming forces of industrial capitalism. Even where social movements are active, as in affluent Western democracies, the power of capitalist interests is well entrenched and likely to militate against the cause of environmental protection. When environmental social movement activity is weak or absent, anti-environmental interests will continue their environmentally destructive practices unchecked.

World Society Theory

The world society perspective explains global environmentalism as driven by the institutional structures and cultural meanings embedded within the international community. A global environmental regime has emerged in past decades, consisting of international treaties, organizations, and cultural understandings regarding the environment that have become accepted and increasingly taken for granted (Frank, Robinson, and Olesen 2011; J. Meyer et al. 1997a; Boli 2006). In contrast to perspectives that view governments as rational actors pursuing their economic or environmental interests, world society theory argues instead that external global forces powerfully shape governments by providing cultural models, scripts, norms, and even identities. Moreover, this book argues that the establishment of global institutional structures creates and empowers environmental "agents" that seek to address a wide range of environmental ills. Drawing on the imagery of the Bee Swarm model, many of these global influences are weak. Improved environmental outcomes result from the aggregation of myriad weak influences, even as many specific treaties, regulatory structures, or social movements may be regarded as failures. Figure 1.5 provides an outline of the world society argument.

Scholars in the world society tradition have examined macro-historical changes in many substantive areas, including education, human rights, women's rights, political democratization, scientific expansion, economic development policies and planning, citizenship, and organizational practices (C. Beck 2011; Boli, Ramirez, and Meyer 1985; Boyle 2002; Castilla 2009; Chabbott 2003; Dimaggio and Powell 1983; Dobbin 2009; Drori, Meyer, and Hwang 2006; Drori et al. 2003; Frank and Gabler 2006; Frank, Hardinge, and Correa 2009; Frank and Meyer 2007; Hironaka 2005; Jang 2000; Kamens and Benavot 1991; Lim and Tsutsui 2011; J. Meyer, Ramirez, and Soysal 1992; Ramirez et al. 1997; Schofer and Meyer 2005; Soysal 1994; Tsutsui and Wotipka 2004; Wotipka and Tsutsui 2008). However, the literature largely focuses on state policies and laws. A growing number of studies look at actual practices or empirical conditions on the ground, such as changes in human rights practices, direct measures of the status of women, and actual degradation of the

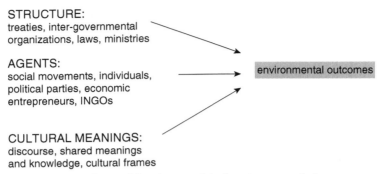

STRUCTURE:
treaties, inter-governmental
organizations, laws, ministries

AGENTS:
social movements, individuals,
political parties, economic
entrepreneurs, INGOs

environmental outcomes

CULTURAL MEANINGS:
discourse, shared meanings
and knowledge, cultural frames

FIGURE 1.5. The world society model of environmental change.

natural environment (Cole and Ramirez 2013; Schofer and Hironaka 2005; Hafner-Burton and Tsutsui 2005; Shorette 2012). This book seeks to extend world society theory by examining when global processes are likely to yield substantive social change.

World society theory is rooted in social constructionist theories (Berger and Luckmann 1967). The power of world society derives from institutionalized culture – shared ideas that underlie our interpretation of reality. Recalcitrant individuals or corporations are not coerced into toeing an environmental line. Rather, their identities and interests are reconstituted as new views on the environment come to be seen as self-evident. Perceptions of empirical reality shift. A perfectly ordinary apple becomes an object of suspicion without the affirming label of "organic." Formerly obscure concepts and issues – BPA-free water bottles, dolphin safe tuna, genetically modified organisms – make their way into routine economic, political, and social discourses. Cultural frames for perceiving and interpreting environmental degradation and injustice are everywhere. New avenues of action become available as well. Where once people might complain about the pollution in a creek, today a wide variety of bureaucratic and legal mechanisms become available to those wishing to bring about improvement. Institutional processes ultimately bring about environmental change to the extent that they reconfigure the social world of individuals and organizations on the ground in various parts of the globe.

The fundamental starting point of the world society perspective is the assumption that actors do not exist *a priori* but are constructed and shaped by social context. Drawing on the insights of Charles Horton Cooley and George Herbert Mead, identities are constructed in dialogue with social interactions (Cooley 1998; Mead 1934). In contrast to modernization and social movement theories, environmentally concerned actors are not assumed to spring into social life fully formed. Instead, the general availability of concepts, language, and social meanings regarding environmental protection enable the construction of environmental actors. This perspective does not deny the occasional pioneer of ecological awareness such as John Muir or Rachel Carson. Yet in order to effect widespread social change, environmentally concerned individuals need institutional structures. Consequently, social change generally results from mundane institutional activities rather than charismatic individuals or revolutionary action.

In a nutshell, social change is driven by the development of institutional structures. In the international environmental arena, this includes the creation of intergovernmental organizations such as the United Nations Environment Programme as well as the growth of international environmental treaties, conferences, and organizations. At the national level, institutional structure is embodied in state legislation, environmental ministries, and government bodies formally tasked with environmental protection. These institutions also embody a set of cultural meanings of the environment that become available across the globe. And institutions empower agents that work to improve environmental conditions.

The Bee Swarm model of change implies that any particular institutional structure may be only weakly linked to substantive improvements in environmental practices. Nevertheless, institutional structures play an essential role in promoting environmental change, through the *creation of a workspace* for addressing environmental issues, the *construction and empowerment of agents* to address environmental problems, and by contributing to *the formulation of cultural meanings*. In a fundamental way, institutional structures establish the workspace in which environmental problems are addressed by setting the public agenda, creating a forum for dealing with environmental concerns, and by codifying formal standards of behavior to which organizations and governments may be held accountable.

Moreover, institutional structures empower an army of agents who work on behalf of social problems. These agents include government bureaucrats, citizens, social movement groups, political parties, economic entrepreneurs, and a wide range of other roles. In contrast to social movement arguments, which often characterize social movements as the starting point of social change, the world society perspective posits that movements are greatly enabled and empowered, and even fundamentally constituted, by institutional structures.

Additionally, the efforts of these environmental agents and institutional structures establish or bolster cultural meanings. The social meaning of the environment changes, encouraging broad awareness and knowledge about environmental issues, the proliferation of environmental identities, and expansive expectations about environmental protection. Over time, even corporate interests shift with the changing institutional structure and expansion of environmental cultural meanings. For instance, corporations are increasingly scrutinized and rated in terms of environmental performance, while consumer tastes have increasingly shifted in ways that advantage "green" companies. As a consequence, economic interests have also moved toward an increasingly pro-environmental stance (Schofer and Granados 2006; A. Hoffman 2001).

In addition to explaining the historical expansion of environmentalism, the world society perspective can help account for variation in pro-environmental change across issue areas. International institutional structures, agents, and cultural meanings have been expanded around some environmental issues and not others – leading to tangible differences in environmental outcomes. Of course, improvement does not necessarily imply the full amelioration of an environmental problem, since processes encouraging environmental degradation are strongly institutionalized as well. Nevertheless, large-scale empirical studies on environmental outcomes have shown moderating effects of institutional structures and environmental agents on the reduction of environmental degradation (See Appendix, also Shorette 2012; Hadler and Haller 2011; Shandra et al. 2009a; Shandra et al. 2009b; Shandra et al. 2008; Jorgenson 2007, 2006, 2003; Schofer and Hironaka 2005). This book provides a full-bodied explanation for the results of these quantitative models from the world society perspective.

Theoretical Issues in the World Society Perspective

One of the remarkable aspects of the world society paradigm is that it offers a perspective that diverges from everyday intuitions. This perspective is exceedingly useful for social analysis, shedding light on processes that are puzzling or simply overlooked by other theoretical traditions. But the non-intuitiveness of world society theory simultaneously makes it challenging to explain. This section addresses five issues that are often raised about the world society perspective: (1) the question of institutional origins, (2) the effectiveness of institutional structures, (3) the role of agency, (4) interests that oppose social change, and (5) conflict among institutional logics. These issues provide the focus for subsequent chapters, so this section also serves as a brief outline of the book. In addition, the theoretical counterfactual of lack of social change, and research implications of the argument, provide the final two chapters.

Institutional Origins and Change

From the world society perspective, social change is driven by the development and expansion of institutional structures in the international sphere. Where do these structures come from in the first place? And, how do they change? Accounts in other scholarly traditions emphasize exogenous forces as the primary motivators of change, in which external geopolitical dynamics, economic interests, political actors, or social movements reshape international institutions. Others look to internal organizational cultures or political struggles within organizations such as the United Nations or World Bank as the source of change.

By contrast, this book develops a more contingent and contextual argument, whereby preexisting international institutions create space for the creation of new international agendas and institutions. Chapter 2 focuses on the emergence of the global environmental regime based on a new cultural conception of the environment rooted in a new United Nations agency. Earlier institutional structures provided the basis for this emergent framework. The new global institution drew cognitive components from earlier global structures such as national parks, bird and animal protection, and resource management policies (Frank 1999; Nash 1973; Hayes 1959). These were combined with cognitive and organizational components from global development institutions and components derived from the United Nations system, resulting in a new framework for the global environmental regime. These prior institutions and cultural frames provided the ingredients from which new institutions were shaped but did not wholly determine the outcome.

Structures and Workspaces

Much of world society research has focused on the worldwide diffusion of institutional structures. The main finding has been the strikingly broad diffusion of policy structures such as legislation, treaties, or governmental organizations (J. Meyer et al. 1997a). Yet a substantial body of scholarship has shown that policies often fail to directly improve outcomes, due to a lack of compliance, resources, or political will (G. Downs, Rocke, and Barsoom 2001[1993]; Chayes and Chayes 2001[1993]). These find-ings argue against the Smoking Gun model of change, in which the enact-ment of the proper law leads to subsequent improvement in outcomes.

As argued in Chapter 3, the lack of effectiveness of policy structures is hardly surprising from the viewpoint of the Bee Swarm model. From the world society perspective, the creation of structure is the starting point rather than the finish line. Rather than enforcing a previously identified solution, the critical role of policy structures is the construction of a workspace in which to figure out solutions. The creation of institutional structures, such as the passage of a new law, officially places a social issue on the political agenda. Institutional structures also provide forums that bring together a variety of agents to work on an issue. In addition, policy structures often codify formal requirements that may mandate accept-able levels of pollutants or specify required pro-environmental behaviors. These effects of the institutional workspace are likely to be indirect, hap-hazard, and may not even appear to be in direct response to the policy. However, the provision of the workspace offered by institutional struc-tures provides an essential first step toward social change.

Actors and Agency

Environmental accounts in the existing literature typically posit that actors provide the central motivation for social change. In modernization accounts, key figures such as Rachel Carson played an essential role in the creation of citizen awareness of environmental problems and the gal-vanization of action. Social movement scholarship also focuses on savvy political entrepreneurs or important social movement organizations that agitate for change. Likewise, corporate actors are central to Marxist accounts of efforts to oppose environmental reform.

The world society perspective, in contrast, theorizes actors as agents constructed within broader institutional structures, as discussed in Chapter 4. Global pro-environmental institutions create and empower a multitude of agents who diffuse and implement policies on behalf of envi-ronmental institutions, oftentimes adapting or innovating in the process. The word "agents" highlights the extent to which pro-environmental individuals and organizations are creatures of wider institutional struc-tures, rather than "actors" with highly autonomous agendas. In particular, social movement activity is characterized as empowered by institutional structures – with the implication that the impact and efficacy of social movements derives substantially from world society. Countervailing institutions that promote economic or political agendas contrary to environmental goals also empower agents, creating much potential political action and conflict.

Interests

In many scholarly perspectives, interests are theorized as centrally important sources of resistance to environmentalism. Efforts to improve environmental protection are likely to be blocked by powerful economic interests and the institutions that support them. Conflict arises when pro-environmental groups promote social change that encroaches on interests of powerful political or economic actors. Moreover, anti-environmental groups frequently enjoy structural advantages in strategic influence and resources, allowing the triumph of anti-environmental interests. From a social movement perspective, corporate interests provide a central reason why policies are ineffective and substantive environmental improvements fail to occur.

World society scholars, on the other hand, have tended to de-emphasize the centrality of material interests. Interests are seen as highly constructed and are rooted in the institutions, laws, and regulations that underlie modern capitalism. In some cases, conflicting institutional arrangements generate conflict. As these conditions shift with the enactment of environmental regulations and the development of environmental culture, corporations easily reconfigure their interests. The implications for this world society perspective on interests are explored further in Chapter 5.

Conflict

Conflict between institutions occurs frequently. Historically, the global environmental institution has collided with the logics of other institutions on several points, such as economic development or national sovereignty. These conflicts have generally been assumed to detract from improvement in environmental outcomes. In the case of deforestation, for example, environmental public goods of expanded forests conflict with institutionalized logics of national sovereignty and economic development. When a single case of conflict is examined at a particular point in time, scholars jump to the conclusion that conflict between institutions has decreased pro-environmental protection.

However, from a broader world society perspective, institutional conflicts might be understood as a necessary consequence of the growth and expansion of environmental institutions as they encroach on the boundaries of competing institutions. Chapter 5 examines the global climate change regime, which currently appears stymied by potent transnational corporate interests. However, from a field-level perspective, global environmental institutions can be seen as audaciously seeking to drastically regulate industrialization. It is hardly surprising that these attempts have met with substantial conflict. Instead, the astonishing part is that global environmental institutions have grown so powerful in only four decades that they now have the ability to challenge fundamental aspects of capitalism.

When Efforts at Social Change Fail

World society can be a powerful source of change. Yet changes, including many that would substantially improve the human condition, sometimes fail to materialize. Chapter 6 draws out the logical implications of previous chapters to examine empirical cases in which pro-environmental change does not come about. In a nutshell, social change is most likely to occur when institutional structures exist and expand, and change is less likely in their absence. Moreover, institutional structures do not always spawn a Bee Swarm of activity. In some cases, institutions fail to empower or mobilize a host of agents. In other cases, shared cultural meanings fail to emerge. These arguments are distinct from the conventional view that environmental efforts fail due to entrenched interests (an argument addressed at length in Chapter 5).

There is a great deal of contingency in world society generally, and in Bee Swarm processes in particular. World society is a diffuse and loosely organized sphere of activity, frequently swept by fads and movement-like swings in activity and attention. Some international conferences or treaties fizzle, while others garner surprising amounts of attention and become watershed events (see Chapter 2). In still other cases, issues simmer for long periods of time before the Bee Swarm suddenly takes off. The arguments developed here are probabilistic in nature. International institutions often matter, but they are not inevitable juggernauts, given the loosely coupled character of world society.

Reflections and Research Implications

Chapter 7 situates this book within the larger trajectory of world society scholarship over the past few decades. Early world society scholarship focused heavily on loose coupling. Characterizing organizational life as "myth and ceremony" helped counter the conventional view that isomorphic change was the result of rational, strategic action. Loose coupling provided an *entre* to argue that culture mattered for policy diffusion; however, it ultimately gave the impression that world society theory could only explain "myth and ceremony" but not actual social change on the ground. This book comes full circle, arguing that world society is indeed a potent source of tangible social change, despite endemic loose coupling across the global system.

The conclusion also provides a summary of arguments in the book and suggests new directions for future research. The Bee Swarm model, for instance, suggests the need for distinctive research strategies for identifying the impact of world society on social activity around the world. In addition, building on Chapter 2, this project offers a template for analyzing the emergence of new institutions in world society. The chapter offers some general theoretical reflections and clarifications, addressing some areas of ambiguity and misunderstanding, in order to refine and amplify world society theory. A final appendix provides statistical models on the effects of environmental institutions and agents on outcomes such as air pollution, water pollution, and the protection of natural areas and forests.

2

The Origins of the Global Environmental Regime

World society theory argues that international institutions propel global social change. This prompts the prior question: Where do international institutions come from? Previous world society research on the environmental regime has looked at broad historical factors that provided a fertile context for new institutional structures to emerge, including the growth of modern science and the organizational expansion of the United Nations system (J. Meyer et al. 1997b; Frank et al. 2000a). This chapter takes a finer-grained look at the process of institutional emergence, seeking to explain the origins of the specific cultural understandings that underlie the contemporary environmental regime. This account focuses on the foundational event of the 1972 United Nations Stockholm Conference and the subsequent establishment of the United Nations Environment Programme (UNEP).

Scholars generally agree that the 1972 United Nations Stockholm Conference on the Human Environment and the subsequent creation of the United Nations Environment Programme provided the foundation for the modern global environmental regime (Rowland 1973; Brenton 1994; Caldwell 1984; McCormick 1995; J. Meyer et al. 1997b). As one environmental scholar proclaims, "The Stockholm conference was the single most influential event in the evolution of the global environmental movement, and of a global environmental consciousness" (McCormick 1995, 127). The Conference, including its planning and aftermath, reformulated the international agenda and spurred the formation of the United Nations Environment Programme.

The Stockholm Conference and the prior years of planning and preparation did more than consolidate preexisting environmental initiatives.

The Stockholm era represents an episode of social construction that fundamentally changed the way that modern environmental problems were understood. A new conception of the environment linked a formerly diverse set of issues under a common umbrella and reframed them as global concerns. This cultural innovation deeply shaped the subsequent trajectory of the global environmental regime.

The post-Stockholm environmental regime reflected systematic constraints imposed by preexisting institutional structures as well as contingent historical events. The Stockholm regime was neither inevitable nor a direct extrapolation from earlier environmental institutions and treaties. International institutions are heterogeneous and loosely organized, in ways that afford space for contingency and the emergence of new forms and configurations. Moreover, institutions empower various participants and constituencies, referred to as "agents" in Chapter 4, who innovate and address new problems. Considerable ingenuity on the part of agents may be involved for the assemblage of diverse institutional components into novel frameworks.

The cultural conception of the environment that emerged from the 1972 UN Stockholm Conference was not created out of thin air. Cognitive and discursive components were drawn from earlier historical discourses of previous international environmental institutions. Moreover, seemingly oppositional global discourses regarding economic development were incorporated and reconciled within the Stockholm framework. Finally, organizational features of the United Nations system and emergent dynamics during and after the conference itself played important roles in forging a new conception of the natural environment.

This chapter first discusses the cognitive framework undergirding the global environmental regime that was constructed in the context of the 1972 UN Stockholm Conference. Next, the conventional explanations offered by modernization and social movement theories are discussed in greater detail. Finally, the chapter outlines a world society account, describing how the content and structure of the global environmental regime developed out of prior institutional structures.

The Modern Conception of the Environment

Today it is taken for granted that the global environmental agenda encompasses a diverse set of concerns including air pollution, ozone depletion in the stratosphere, water pollution, land use, waste disposal, biodiversity, deforestation, desertification, and overuse of marine resources

(Tolba and El-Kholy 1992). Moreover, these issues are routinely perceived as global in scope. However, the underlying cognitive framework that binds these issues together is quite recent, and clearly linked to the 1972 UN Stockholm Conference. As one environmental scholar notes, the Stockholm Conference exemplified the shift "from the limited aims of nature protection and natural resource conservation to the more comprehensive view of human mismanagement of the biosphere" (McCormick 1995, 128). This section describes the conception of the environment that has become the cognitive basis for the expanding global environmental regime.

Pinpointing the specific origin of a particular conceptual framework within the flow of history is a challenge for historical scholarship. There is a strong temptation to read contemporary meanings into the events of the past. It is all too easy to assume that efforts to create natural parks or to protest oil spills in the early half of the twentieth century sprang from the same understandings of nature that would motivate such actions today. However, earlier historical efforts at environmental protection were not informed by the ecosystemic perspective that is conventional today, in which nature is conceived as a deeply interrelated whole. Instead, international treaties and national legislation historically parceled out issues into separate policy domains (Nash 1973; Hayes 1959; McCormick 1995). Only relatively recently have such issues become viewed as related facets of a general environmental agenda.

Another challenge of identifying the cognitive origins of institutional structures is that they are built from prior institutional structures. The intellectual roots of a new structure can be discerned in earlier institutions, especially when viewed in hindsight. In the case of the environment, biological scientists and natural philosophers had previously discussed the physical interconnectedness of the natural environment. Yet according to the Oxford English dictionary, the contemporary meaning of the words "ecology" and "environment" are quite recent constructions. In 1963, Aldous Huxley was the first to use the word "ecology" in the modern context as signifying the interrelationship between living creatures and the physical environment in a paper titled "The Politics of Ecology" (Brenton 1994, 15). Previously, the word "ecology" had been understood purely as a scientific term.

International efforts prior to 1970 addressed aspects of nature as a series of discrete and separate domains, each with its own institutional structure. Migratory birds, Mediterranean fish, and Pacific whales were each protected under a different international treaty structure and appealed to diverse constituencies (McCormick 1995). Similar fragmentation can be seen at the national level. In the United States, air pollution was historically the responsibility of the Secretary of Health, Education, and Welfare; water pollution was originally under the authority of the Surgeon General of the Public Health Service, and jurisdiction for wilderness protection was divided among the U.S. National Park Service, the U.S. Forest Service, the U.S. Fish and Wildlife Service, and the Bureau of Land Management (R. Andrews 2006).

Related to this, environmental problems were historically perceived as local issues. The smog in Los Angeles, London, Bombay, and Tokyo were traditionally understood as separate and local urban concerns. It took a new cultural framework to view them as specific instantiations of the broader phenomenon of planetary air pollution. An oil spill off the Santa Barbara beach in 1969 was not perceived as linked to the *Torrey Canyon* oil tanker disaster in 1967 that threatened the British coast. A different imagination was needed to view both as potentially threatening the interrelated links of the global marine ecosystem. The discourses and cultural frames emerging from the Stockholm Conference reimagined disturbances in remote parts of the environment as interlinked and potentially disruptive to components of the ecosystem of more direct concern to humans. Based on this contemporary vision of the ecosystem, "issues may be localized in particular countries but impact upon many others, and are thus amenable to international policy agendas. And these problems may have implications extending beyond the particular countries to the entire world, for example, tropical deforestation" (Caldwell 1984:12–13).

The conception of nature as an interrelated whole of global concern was in large part an accomplishment of the 1972 UN Stockholm Conference (Caldwell 1984; Rowland 1973). This contemporary understanding of the ecosystem emphasizes the interrelatedness of humans with the natural environment and the interconnection among physical and biological processes on a planetary level. The scope of this cognitive framework encouraged the creation of a comprehensive global agenda linking an expansive array of issues previously seen as unrelated. As a result, it is now taken for granted that the broad sweep of environmental concerns encompasses human influences on air, water, land, and sea. Maurice F. Strong, chair of the 1972 UN Stockholm Conference, retrospectively claimed that the conference had brought forth:

... the realization that the environmental issues are inextricably linked with all other factors in contemporary world politics; that we urgently require not only a new perception of man's relationship with the natural world, but with man's relationship to man; that the problems of the rich cannot be seen in isolation from those of the poor; that in all respects we inhabit Only One Earth. (Rowland 1973, ix)

The strength of this expansive vision of the environment stemmed from the diverse institutional components that went into its creation, as argued in this chapter.

This modern conception of the environment has become so widely institutionalized today that it is easy to think that it was generally self-evident in the late 1960s. Both modernization and social movement perspectives assume that the broad planetary scope of environmental issues was self-apparent. In contrast, the world society perspective argues that this broad conceptualization was an accomplishment of agents working in the international sphere and building on prior international institutions. Institutional structures were required to bring widespread recognition of environmental problems in the first place. Institutional structures provided the necessary resources for the construction of scientific consensus on environmental problems. Finally, international institutions supported the global mobilization of effective social movement groups.

Traditional Perspectives on Institutional Formation

The traditional literature on the origins of the global environmental regime largely bypasses the formation of cognitive conceptions of the environment. Scholars typically assume that environmental degradation was broadly evident to citizens and policymakers, or at least that scientific knowledge translated straightforwardly to public knowledge. In particular, both modernization and social movement perspectives assume that the severity of environmental problems was recognizable by both governments and concerned citizens; the perspectives differ primarily in the responsiveness of governments to these problems. This section outlines modernization and social movement explanations for the origins of the global environmental regime.

Modernization Theory

Modernization theory explains the origins of contemporary environmental institutions as a response to the deterioration of the natural environment. Scientists or other experts may mediate the recognition of environmental degradation for issues less apparent to laypeople, such as the identification of problems in the earth's ozone layer, or the development of theories of the greenhouse effect on climate (E. Haas et al. 1977; P. Hass 1992; Bocking 2004). Yet from the modernization perspective, objective reality itself is the primary motivator for the emergence of national and international environmental regulation. By contrast, the world society perspective argues that environmental realities and academic science are an insufficient basis for the creation of global institutions. Instead, the reverse is typically the case: international structures provide the resources to fund and coordinate scientific research of the scope needed to identify and construct new environmental problems.

Environmental histories often begin with a "creek of my youth" story that recounts an idyllic childhood brook now filled with dead fish and old tires (Bocking 2004). Such accounts suggest that the basic forms of environmental degradation were easily observable. These accounts are dubious as an explanation for the modern environmental regime. Many of the most obvious sources of environmental degradation had improved in the industrialized countries by the late 1960s, as the result of urban planning policies and other precursors of the modern environmental regime. For instance in the United States, levels of air pollutants such as carbon monoxide and sulfur dioxide had dramatically decreased, reducing smog levels in cities such as Los Angeles. Similarly, the historically famous fogs of London were disappearing due to an 80 percent reduction in smoke and a 40 percent decrease in sulfur dioxide emissions (Brenton 1994, 21). Rivers such as the Thames and the Columbia were being revived. As Lynton Caldwell noted, by the early 1970s, "environmental conditions were better [in some respects] than they had been a generation earlier: city streets were cleaner, water-borne disease less common, bituminous coal smoke banned or reduced in the atmosphere" (1984, 26).

A more sophisticated explanation is that the global environmental regime was motivated by scientific discoveries that identified forms of environmental degradation that were not easily observable by laypeople (P. Haas 1992; E. Haas et al. 1977). Scholars have argued that the development of a scientific consensus about the severity of an ecological problem – an epistemic community – is often a critical first step in institutional formation. This scientific agreement on the identification of an objective environmental issue provides a solid basis on which to develop policy structures and institutions.

One familiar account claims the 1962 publication of *Silent Spring* by biologist Rachel Carson increased public awareness of the damage wrought by pesticides and provided the impetus for pro-environmental movements and institutional structures (but see D. Meyer and Rohlinger

2012). A similar case is the scientific discovery of the reaction between chlorofluorocarbons and ozone molecules leading to the thinning of the earth's ozone layer, which is discussed in Chapter 3. In these examples, science is perceived as playing the pivotal role in the initial identification of an environmental problem that motivates public action.

Yet a large body of scholarship has noted that the relationship between research science and international politics is not a comfortable one (Andresen 1989; Andresen et al. 2000; Dimitrov 2006; Jasanoff 1990; Lidskog and Sundqvist 2002; Litfin 1994; Miller and Edwards 2001; Parson 2003). Academic scientists are typically wary of over-generalizing a laboratory result to the scope necessary for international policymaking. Nor are scientists adept at publicizing their research in terms easily comprehensible to politicians and the public. Moreover, argument and controversy are far more common in academic scientific discourses than is consensus on the magnitude, or even the existence, of an environmental problem. These difficulties should not be taken as implying that scientists bias their research to pander to politicians or that ecological problems are not empirically observable phenomena. Yet scholarship on the sociology of science has shown that the formation of scientific consensus rarely occurs in an easy or straightforward manner.

Indeed, scholars have observed that environmental policies have in many cases been implemented despite a fairly high degree of uncertainty or controversy in scientific findings (Andresen 1989; Andresen et al. 2000; Dimitrov 2006; Harrison and Bryner 2004; Jasanoff 1990; Lidskog and Sundqvist 2002; Litfin 1994; Miller and Edwards 2001; Parson 2003). One survey of the literature finds that "significant scientific uncertainties and gaps of knowledge about the problem at hand remained" at the time policy commitments were made (Dimitrov 2006, 2). This lack of reasonable scientific consensus has led to the adoption of the "precautionary principle" in international law that postulates preventive legal action may be mandated even if scientific consensus on an environmental threat has not yet been achieved (Dimitrov 2006).

In contrast, the world society perspective suggests that the role of science may be the reverse of that posited by modernization theory – institutional structures and policymaking processes enable the undertaking of scientific research of the magnitude necessary to speak to policy issues. Original laboratory findings from a single scientific team may provide the first step in the identification of environmental damage. However this rarely supports generalization of the kind needed for global policymaking. Policymaking requires institutional structures that coordinate teams of scientists from different countries, that supply reams of data from industries or governments that would otherwise be inaccessible, and of course to provide funds for extensive scientific activities. For instance, in the case of the ozone depletion regime, several national governments and international organizations including the Organisation for Economic Co-Operation and Development (OECD) and the United Nations Environment Programme provided the institutional resources that were needed to expand the findings of initial studies into a broad scientific consensus that became the basis for a worldwide treaty on chlorofluoro-carbon production, as discussed in Chapter 3.

Science does play an important role in the formulation of environmental problems. However, scientific research alone cannot provide sufficient impetus for the creation of global environmental institutions. Instead, the world society perspective suggests that science typically provides a post hoc rationalization for policymaking, rather than the initial impetus. As one sociologist of science concludes, "A review of most of the international treaties negotiated since the 1972 Stockholm conference shows that scientific evidence has played a surprisingly small role in issue definition, fact-finding, bargaining, and regime strengthening" (Andresen et al. 2000, 1). In sum, objective knowledge about environmental degradation is not single-handedly responsible for the construction of the modern global environmental regime.

Social Movement Accounts

A second explanation for the origins of the global environmental regime comes out of the social movement literature. Scholars have argued that burgeoning social movements in several Western countries spurred international recognition of environmental problems, eventually leading to the 1972 UN Stockholm Conference (Caldwell 1984; Brenton 1994; McCormick 1999). However, social movement accounts fail to appreciate the difficulties of bringing together social movement groups from multiple countries that were often focused on different issues and employed diverse strategies. The world society perspective suggests the reverse: the formation of the global environmental regime created conditions that enabled social movement activity by creating new frameworks for understanding environmental problems, legitimating environmental issues on national and global agendas, and by providing a forum that brought together citizen groups and activists from countries all over the world.

Histories of the international environmental regime often begin by noting the growing concern over environmental issues in the 1960s in a few Western industrialized countries (Caldwell 1984; Brenton 1994; R. Andrews 2006). For instance, the first Earth Day in the United States in 1971 was the largest part-celebration, part–social protest event that had occurred up to that time (McCormick 1995, 55). In the United States, the number of survey respondents who believed pollution was among the most important of problems for the government quadrupled between 1965 and 1970 (Brenton 1994, 19). Public concern for the environment also increased in several Western European countries including France, Germany, the United Kingdom, Sweden, Switzerland, Japan, and the Netherlands (Brenton 1994; Kitschelt 1993; Broadbent 1998). Caldwell claims that, "During the 1960s a crescendo of public concern, first expressed in the popular press, culminated by the end of the decade in national environmental laws and policies" (Caldwell 1984, 25–26).

However, the environmental social movements of the 1960s were quite local in their orientations and conceptions of environmental protection. Rather than banding together under a broad umbrella that united diverse ecological concerns, social movements typically focused on narrow issues of local concern. The 1969 oil spill in Santa Barbara, California aroused local social movement activism that was unconnected with protests against the 1962 proposal to build a hydroelectric plant in the Storm King Mountains of New York or the 1968 protests in Japan over the proposed building of Narita Airport in Tokyo. Those disparate issues were not yet seen as falling within a general framework of protecting the planetary ecosystem.

Secondly, there is little to suggest that these social movements had influence on the world stage. A major obstacle to the transmission of local social movement activity to the international level has been the difficulty of coordinating multiple citizen groups around the world. As argued in Chapter 4, institutional structures generally precede effective social movement action in part because they provide the organizational apparatus needed to coordinate action. Before the 1972 UN Stockholm Conference forged a broader conception of the environment as an umbrella issue for disparate environmental issues, activists had difficulty identifying groups with similar agendas in other countries, much less coordinating networks among groups amidst different cultures and languages.

Finally, domestic social movements were limited to a very small number of countries. This is a critical problem for the creation of a successful international regime, since such a regime generally requires the participation of a broad spectrum of countries. The governments and citizens of most developing countries had little interest in the various antipollution movements emerging in industrialized societies, instead focusing on issues such as development, poverty, and the development of basic governance following independence. Scholars noted rare instances of environmental protests in the global south, such as India and Mexico, cautioning that they "certainly did not point to the sort of generalized environmental alarm which, in the West, forced governments to give the issue ... political priority" (Brenton 1994, 31). Nevertheless, developing nations represented the majority of votes within the United Nations assembly and their support was essential for the creation of UNEP and broad-based environmental treaties.

World society theory suggests the converse of the social movement hypothesis. The emergence of institutional structures such as UNEP was critical for the development of citizen awareness and the mobilization of protest groups in disparate countries. The Stockholm Conference explicitly established an Environment Forum with the specific goal of encouraging the participation of non-state organizations, citizens, and activist groups. The Environment Forum was an official part of the Stockholm Conference and was attended by representatives of approximately 400 non-governmental organizations (NGOs) (de Lupis 1989, 212). Additional forums, such as Alternative City, the Oi Group, and the Hog Farm, sprang up semi-officially around the Stockholm Conference. These forums built on existing environmental citizen groups and also motivated the formation of new ones (Brenton 1994). While most scholars doubt that these forums influenced the substance of the proceedings at the official Stockholm Conference, these citizen forums did play a role in constructing social movement groups and agendas focused on both international and domestic environmental issues (Caldwell 1984; Schechter 2005; McCormick 1995).

In sum, the link between social movements and the emergence of national and especially global environmental institutions is tenuous at best. Early environmental mobilization was fragmented and quite local in its orientation. While social movement mobilization in a handful of Western countries was part of the larger social milieu from which Stockholm and UNEP emerged, there were few direct connections linking them. Instead, the emergence of a global regime was highly consequential for social movements by providing a broader frame to link previously disparate activists, placing environmental issues on the global agenda, and constructing spaces and opportunities (including the Stockholm Conference itself) for national groups to connect and mobilize a global environmental movement.

World Society Process of Institutional Formation

This chapter argues that new institutional structures are built on the foundation of prior structures. However, the particular configuration of components and the resulting structure are not necessarily a straightforward extrapolation of prior institutions. As discussed in Chapter 4, institutional agents interpret, translate, and adapt existing institutional components into new institutional forms. In the case of the global environmental regime, the unique blend of institutional components enabled a broad framing of environmental protection and encouraged the subsequent expansion of the international environmental regime.

This section examines preexisting institutional structures that were absorbed into the post-Stockholm cognitive framework for the global environmental regime. The Stockholm Conference built on early international efforts at managing the environment and natural resources, each of which had its own set of treaties, organizations, and constituents. Components were also drawn from the wholly different sphere of the global regime of activities addressing economic development. Finally, the United Nations system contributed its own bureaucratic constraints as well as organizational mechanisms that were consequential for the emergence of the environmental regime. The resulting new institution emerged as more than the sum of these preexisting structural parts.

Preexisting International Environmental Institutions

The cognitive conception of the environment that emerged from the Stockholm Conference drew on three different frameworks for environmental protection. In the 1960s, these prior structures were institutionalized in the international sphere to varying degrees, with their own organizations, discourses, and constituencies. Each served as the basis for multiple international conferences, any one of which had the potential to develop into a full-blown international regime rooted in a distinctive conceptualization of the environment. As it happened, the Stockholm initiative drew on and integrated these cognitive ingredients in a generative way, paving the way for the expansive modern environmental regime familiar to us today.

One set of early international efforts employed a "preservation" framework that focused on the protection of beautiful, pristine, or spectacular aspects of nature (McCormick 1995; Nash 1973). Originating in the late nineteenth century in the United States and the United Kingdom, the preservationist movement traced its initial foundation to the writings of John Muir and the formation of the national park system in the United States. In Europe, early efforts drew from the Romantic poets and the heath preservation societies of the United Kingdom (Nash 1973). Following the preservationist logic, the first international environmental treaties focused on the protection of migratory birds. Other spectacular aspects of nature, such as exotic African animals and areas with recreational utility for humans, were also deemed fit for international protection (McCormick 1995).

The preservation framework provided the basis for several international conferences going back to the turn of the twentieth century (Caldwell 1984). These included the International Congress for the Protection of Nature (Paris, 1909), the International Congress on the Protection of Flora, Fauna, and Natural Sites and Monuments (Paris, 1923), the International Congress for Study and Protection of Birds (Geneva, 1927), and the Second International Congress for the Protection of Nature (Paris, 1931). With the establishment of the United Nations in 1945, a number of preservation-oriented conferences sprang up in connection with United Nations agencies as well. UNESCO sponsored a conference at Fontainebleau in 1948, following a prior conference held by the Swiss League for the Protection of Nature at Brunnen in 1947 (Caldwell 1984). In 1949, UNESCO sponsored another conference jointly with the International Union for the Protection of Nature (IUCN) that became the International Technical Conference on the Protection of Nature (McCormick 1995).

Although the preservation framework was the basis for the oldest environmental treaties and cooperative efforts, it had features that posed challenges for the development of a broad-based global environmental regime. Most aspects of preservation fell within the bounds of national sovereignty, such as the creation of national parks or the conservation of endangered species, and therefore were outside the ambit of international organizations and treaties. Perhaps more problematic, preservation was based on a distinctive Western discourse that was not widely adopted or appreciated in non-Western states. Indeed, the preservation tradition often involved protecting wild animals from indigenous populations in Africa and Asia so that European hunters could enjoy exotic game (Hayden 1970[1942]). Unsurprisingly, the preservation conferences elicited little interest from countries of the global South. As McCormick noted, preservation conference organizers "realized that nature protection was not the best response to the growing post-war pressure on natural resources; the focus should instead be on promoting conservation as an integral part of development" (McCormick 1995, 48). In the end, proponents of the preservationist framework were unable to generate the broad-based participation needed to form an expansive international regime.

A second framework for early pro-environmental efforts was rooted in "resource management" logics. The resource management perspective focused on the efficient conservation and utilization of natural resources by humans (Nash 1973; Hayes 1959). Human interests were central to this approach, in contrast to the preservation view that nature ought to be protected for its own sake. The resource management frame assumed that humans needed to be protected from their own shortsighted tendencies and encouraged to carefully husband natural resources in order to maintain their maximum sustainable utility. Resource management proponents argued that wise policies were needed to ensure that humans take a long-term perspective for eventual maximum benefit. By the late 1960s, the resource management frame had generated a set of structures at the international level, including treaties to monitor fishing in the Atlantic and whaling in the Pacific (McCormick 1995).

The resource management framework also spawned a set of international conferences. The United Nations regional organization ECOSOC sponsored a conference in 1949 titled the United Nations Scientific Conference on the Conservation and Utilization of Resources (UNSCCUR) (McCormick 1995; Rowland 1973). In contrast to the handful of Western states that participated in the preservation conferences, UNSCCUR was attended by 49 countries as well as FAO, UNESCO, the World Health Organization, and the International Labour Organization (McCormick 1995, 40). In addition, a number of environmental conferences were held by United Nations agencies throughout the 1960s, including a conference on new energy sources at Rome in 1961 and four on the peaceful uses of atomic energy in Geneva in 1955, 1958, 1964, and 1971 (Caldwell 1984, 39). Another United Nations conference was held in Geneva in 1963 on the Application of Science and Technology for the Benefit of the Less-Developed Areas. Immediately prior to the 1972 UN Stockholm Conference, UNESCO sponsored the Biosphere Conference in Paris in 1968 (McCormick 1995, 107). This conference was designed as a "general conference on [the] rational use and conservation of the biosphere" that sought to encourage participation from the global South (McCormick 1995, 108). The Biosphere Conference focused on plant and animal resources, and paid little attention to human environments (Caldwell 1984, 40). The 1972 Stockholm Conference, with its odd formal title of the UN Conference on the Human Environment, was envisioned as a complement to the resource-focused Biosphere Conference.

The resource management framework appeared to have the greatest potential for translation to a broad global institution within the United Nations. The economic sustainability of natural resources was of direct interest to both developing and industrialized states, and the United Nations conferences on resource management had generated broad participation among nation-states. The resource-oriented framework complemented rather than conflicted with national economic development goals and the rapidly growing international development regime. The history of multiple United Nations–sponsored conferences had strengthened and expanded this institutional framework at the international level, making it the frontrunner as the basis for a global environmental institution.

A third early institutional framework addressed pollution and related issues. While pollution was on the national agendas of several industrialized Western states by the late 1960s, there had been only a few pollution-oriented treaties in the international realm. This paucity was due to the perception that pollution was mainly a local issue. Few cases of trans-boundary pollution had been identified. Since pollution was mainly understood as a domestic concern in the late 1960s, it did not provide a substantial basis for international institutions. Nevertheless, some aspects of the pollution framework made it onto the international agenda, primarily focused on issues of marine pollution such as the problem of oil spills outside of territorial waters. For instance, the Technical Conference on Marine Pollution and Its Effects on Living Resources and Fishing had been held in Rome in 1970 under the auspices of the UN Food and Agriculture Organization (Tolba and Rummel-Bulska 1998).

The initial call for what would become the Stockholm Conference was rooted in one of the early instances of transnational pollution. In particular, Sweden had become alarmed about the problem of acid rain, in which sulfur dioxide emissions from its industrialized neighbors (especially the United Kingdom) were drifting over the border to Sweden. Since the problem did not originate on Swedish territory, a transnational organization such as the United Nations was needed to act on the problem. Thus in 1968, the Swedish delegation called for an environmental conference at a meeting of the UN Economic and Social Council (ECOSOC) (Rowland 1973; McCormick 1995).

Although the pollution frame provided the initial impetus for the 1972 Stockholm Conference, it did not appear to be an especially promising or resonant framework. On the one hand, the industrialized countries resisted interference with their industrial interests. On the other hand, the less developed countries did not see pollution as a pressing issue, since it was mainly a problem in industrialized countries. Few envisioned that the 1972 conference would reformulate the scope of environmental problems such that the pollution problems in one location would routinely be seen as affecting the interconnected biological processes on the planet as a whole.

In sum, three international institutional structures, each with its own distinctive discourse and proponents, provided key elements for a broader conceptual framework that was the basis for the post-Stockholm environmental regime. The influence of these early institutional structures was not surprising given their presence on the international agenda. Any one of these frameworks, especially the resource management framework, might have been expanded to form the basis for a global regime. Instead, the new regime was built around an amalgamation of all three. This was a non-trivial accomplishment, as the early discourses were quite distinct at the time. The creation of a unified cognitive framework from these heterogeneous components required significant intellectual effort and entrepreneurship on the part of institutional agents during the conference and its aftermath.

The Economic Development Regime

These nascent pro-environmental efforts were dwarfed by other international agendas in the 1960s, most notably the rapidly expanding international development regime, which sought to encourage industrialization within impoverished nations. The development regime was an obvious major obstacle to the creation of institutional structures addressing environmental protection. Pro-environmental discourses of the era were generally framed as diametrically opposed to the increasingly urgent calls for national industrial development, making it difficult to generate broad international support for pro-environmental treaties and institutions. In the case of the Stockholm Conference, the initial antipollution framework of environmental protection was stretched and reformulated to include issues of economic development that were of primary importance to the developing countries. This innovation – the alignment of sustainable development with environmental protection – had its origins in the Stockholm Conference and has had long-lasting impact on the global environmental regime.

During the 1960s, the countries of the global South were predominantly concerned with issues such as economic development and the challenges of nation-building that followed colonial independence. Economic development for the global South had only recently become institutionalized as a global concern. The rapid pace of decolonization during the 1960s led to swelling numbers of developing countries, and former colonial powers were seen as bearing some responsibility for their economic development. Industrial pollution appeared far removed as a concern for these newly independent and economically underdeveloped states. Consequently, proposals for the creation of a global environmental regime initially met opposition from states in the global South.

The developing states broadly viewed environmental concerns, particularly those from the preservation or pollution frameworks, as indulgences of prosperous Western states that could afford to be concerned about luxuries such as national parks, the conservation of wildlife, or the ills of industrial pollution that underdeveloped states could only dream of suffering. As one UNEP document noted, "Debates on doomsday theories, limits to growth, the population explosion, and the conservation of nature and natural resources … were thought of as largely academic, of no great interest to those faced with the daily realities of poverty, hunger, disease and survival" (quoted in McCormick 1995, 112). Consequently, one scholar observed, "The major developing countries approached the [Stockholm] conference with caution bordering on hostility" (Brenton 1994, 37).

In particular, the antipollution framework proposed for the Stockholm Conference was not exciting to the non-Western members of the UN General Assembly. Far from agreeing with Western claims that pollution was a global problem, leaders of developing states argued instead that the wealthy states that created the pollution should also pay the costs of cleanup (Tolba 1982). Indeed, at the Stockholm Conference "the Ivory Coast announced that it would like to have more pollution problems, 'in so far as they are evidence of industrialization'" (quoted in Rowland 1973, 50). Indira Gandhi of India claimed, "poverty, not pollution, was the principal problem confronting India" (quoted in Brenton 1994, 37). The Brazilian ambassador to the United States summarized the general hostility of the global South to an environmental regime when he wrote in 1971, "Environmental deterioration, as it is currently understood in some developed countries, is a minor localized problem in the developing world" (quoted in Brenton 1994, 39).

Yet the participation of the global South would be necessary for the creation of any pro-environmental institution under the auspices of the United Nations. In order to create a United Nations agency, environmental protection needed to be reformulated to appeal widely to the members of the UN General Assembly. An issue of concern to only a handful of nations, even the politically powerful states of the West, was likely to be relegated to one of the regional bodies of the United Nations. Previous international conferences on the environment had failed to attract much participation from the developing states, as previously discussed, and the same may have occurred in the case of Stockholm if not for innovations by the conference organizers.

The concept of "the pollution of poverty" was developed during one of the preparatory conferences for Stockholm held at Founex. The term "pollution of poverty" was used specifically to describe the type of environmental degradation that often occurred in less developed states (Tolba 1982, 7). As the report summarized, "The central environmental problems facing developing countries ... stem not from pollution but from poverty, disease, hunger, and exposure to natural disasters. The solution to these problems was to be found through the process of economic development itself" (Brenton 1994, 38). More broadly, the concept was used as a bridge that incorporated the concerns of economic development into the cognitive framework of environmental protection.

The emergent cognitive framework of environmental protection shifted with the incorporation of these claims that environmental problems in less developed states "were rooted in social and political problems" (McCormick 1995, 128). These outcries from the global South "led ultimately to a much wider view being taken of the roots and causes of the environmental crisis" (McCormick 1995, 128). The incorporation of economic development discourses resulted in a reconceptualization of the environmental framework that made it more amenable for participation by the states of the global South.

This new formulation proved long lasting, as can be seen in the 1992 United Nations Conference on Environment and Development with its focus on "sustainable development." The tensions between the original pollution framework of the Stockholm Conference and competing institutional logics of economic growth are still evident today, as discussed further in Chapter 5. Had a different cognitive framework become the basis for the global environmental institution, particularly one drawn from a resource management framework, these struggles to align economic development with environmental protection might have been less troublesome. Nevertheless, the incorporation of sustainability to the agenda was to prove an enduring basis for the global environmental institution.

The Workspace of the UN Stockholm Conference

The third important factor shaping the emergence of the global environmental regime was the organizational structure of the United Nations system and the Stockholm Conference itself. One of the central features of institutional structures, including the United Nations, is that they provide a "workspace" in which new activity and innovation may occur (a concept that is developed further in Chapter 3). Chapter 3 argues that institutional structures do not simply implement fully worked-out solutions to social issues. Instead, they set the agenda and bring various constituencies together, providing a workspace in which an issue can be formulated, discussed, and refined, and in which potential solutions can be aired. The United Nations system provided a broad organizational structure for action, while the Stockholm Conference provided the specific workspace for the development of a cognitive framework that was expressed in the formal Declaration of the United Nations Conference on the Human Environment (UNCHE 1972).

The United Nations System

The immediate origins of the Stockholm Conference can be traced to mundane bureaucratic processes within the United Nations. Rather than focusing on heroic social movements or persuasive issue-entrepreneurs, world society scholars highlight the role of mundane and incremental organizational processes that often generate new institutional structures. The global environmental regime arose straightforwardly out of the United Nations conference system. The capacity for a Special Conference was a routine structural mechanism within the United Nations. Assigning the topic of environmental protection to a Special Conference did not call for unusual departures from the bureaucratic workings of the United Nations. However, there was little reason to expect this Special Conference to be more successful in the creation of a new global regime than were previous or subsequent Special Conferences.

Prior to the 1972 Stockholm Conference, the lack of interest displayed by developing countries meant that international environmental issues were more likely to be organized at the regional level rather than by the United Nations. Regional organizations such as the Council of Europe and the Organisation for Economic Co-Operation and Development (OECD) had begun to establish environmental programs during the 1960s. Other regional organizations similarly developed environmental agendas, including the Organisation of American States (OAS), the North

Atlantic Treaty Organization (NATO) and Comecon (CMEA) (de Lupis 1989, 209).

Proposals for a new global environmental institution within the United Nations faced direct competition from preexisting UN agencies. A new UN environmental agency would encroach on the boundaries of at least twenty United Nations agencies that claimed they already had jurisdiction over aspects of the environment (McCormick 1995). Representatives of eight United Nations agencies attended the 1972 Stockholm Conference, including the International Civil Aviation Agency (which claimed jurisdiction over airport noise and sonic booms); the World Meteorological Organization (which laid claim to man-made climate change); the International Atomic Energy Agency, the World Bank, and even the General Agreement on Tariffs and Trade (Rowland 1973).

However, the United Nations provided its own organizational innovation designed to sidestep these sibling rivalries: the United Nations Special Conferences. While the various United Nations agencies held conferences of their own, the Special Conferences were intended to bring new items onto the United Nations agenda (Schechter 2005). The first Special Conference was the 1968 International Conference on Human Rights. The 1972 United Nations Conference on the Human Environment would be the second. Yet while the Special Conferences promised to broaden the agenda of the United Nations, few have actually led to the creation of a new United Nations agency. As Caldwell complains, "The accomplishments of United Nations conferences have generally not been impressive" (1984, 51).

In sum, the proximate origins of the Stockholm Conference arose from the normal bureaucratic workings of the United Nations organizational apparatus. There were no guarantees that the 1972 Stockholm Conference was the event that would become the origin of the global environmental regime. Instead, its success was due to the work of agents that pulled disparate elements from preexisting international institutions into a resonant and coherent framework that could serve as the basis for a global institution with broad-based support in the international community.

Workspace of the Special Conference

The 1972 UN Stockholm Conference provided the particular workspace in which a global environmental agenda was initially drafted. This workspace stretched through the four years of preparation for the conference, through the conference itself, and included efforts to craft a synthesis following the conference. The conference directly resulted in the adoption of the Declaration of the United Nations Conference on the Human Environment (UNCHE 1972). In addition, a philosophical statement written by Barbara Ward and René Dubos (1972), published as *Only One Earth*, was written at the request of the conference. Although the Declaration served as the jumping-off point, post-conference work also helped to synthesize the hodgepodge of issues into a coherent cognitive framework for the nascent global environmental institution.

Preparations for the Stockholm Conference began four years in advance and entailed innumerable trips by conference organizers to every continent except for Antarctica (Rowland 1973). Such preparatory work was especially critical for the formation of a new institution that did not yet have an organizational structure or well-defined agenda. Although environmental issues were not on the political agenda of most countries in the world in the late 1960s, each participating state was asked to write a report on the state of their environment in preparation for the Stockholm Conference. As Caldwell muses, "The consequences of this preconference activity may have had a more extended and lasting significance than did the actual conference itself ... the years 1968 to 1972 witnessed a worldwide raising of consciousness for which there appears to have been no precedent" (1984, 47).

As a result of these preparations, participating states arrived at the 1972 UN Stockholm Conference with a wide array of issues and concerns to be mixed into the pot. Since no global environmental regime yet existed, there was little consensus on what counted as an environmental issue. Consequently, delegations brought a heterogeneous assortment of issues to be discussed. Surprisingly, quite a few issues that would be viewed as irrelevant today were included in the 26 principles and 109 declarations that were agreed upon at the Stockholm Conference. As one scholar notes of the declaration that resulted from the Stockholm Conference, "Such texts tend to become a Christmas tree to which every country endeavors to hang its own pet projects" (Brenton 1994, 46).

One indicator of the heterogeneity of issues discussed at Stockholm was the set of supplementary papers that were submitted by participating states. States were allowed to submit additional material on an issue of interest as well as the required report on the state of the environment in each country. These supplementary papers were particularly farfetched in their topics, yet were diplomatically included in the principles that resulted from the conference. For instance, Peru submitted a paper on mercury contamination in the sea, which along with cadmium poisoning was also a concern for Japan. Accordingly, Recommendation 48b of the UNCHE Declaration noted: "Discharge of toxic chemicals, heavy metals, and other wastes may affect even high-seas resources" (UNCHE 1972). Spain had submitted a paper on forest fires, Australia had submitted a paper on insect viruses that might be used as insecticides, and Japan had included disease-related pollution in its paper. In recognition, Recommendation 26 read: "It is recommended that the Food and Agriculture Organization of the United Nations co-ordinate an international programme for research and exchange of information on forest fires, pests and diseases" (UNCHE 1972).

Similarly, political issues of colonialism, imperialism, and racism were affixed to the environmental agenda (Rowland 1973, 50–51). For example, during the conference the Nigerian delegation denounced the presence of "the racist government of South Africa and other agents of colonial oppression like the Portuguese," arguing that "They cannot treat the environment with concern and consideration if they treat the vast majority of mankind inhabiting the countries which they dominate with less than human consideration" (quoted in Rowland 1973, 51). The Tanzanian delegation declaimed, "The evils of *apartheid*, racial and colonial oppression, far from being irrelevant, are at the very core of environmental problems in Africa" (quoted in Rowland 1973, 52). The Libyan delegation gave a lengthy peroration that decried many current political events including

the domination of minorities on people's abilities and destinies that is happening in several areas of Africa where, as well as in the USA, people are still suffering from racial segregation [and] discrimination. It also cannot ignore destruction of [the] human environment in Indochina by biological warfare and modern equipment and incineration materials. The conference, besides, cannot ignore mass destruction, spoiling lands and crops, environmental pollution by human bodies who are unfairly killed, nor mass murders of moslems minority in [the] Philippines, all these aspects and events put the historical and human responsibilities on the conference to face the reality that "Man" is subject to extermination besides what is caused to this environment of destruction and corruption. (Rowland 1973, 51)

Such sentiments resulted in Principle 1: "… policies promoting or perpetuating apartheid, racial segregation, discrimination, colonial and other forms of oppression and foreign domination stand condemned and must be eliminated" (UNCHE 1972).

Zambia had also submitted a paper on ecological upheavals due to Portuguese colonial administration and the effects of chemical warfare by

South Africa as well as denouncing hydroelectric plans in Mozambique and Angola. These arguments, as well as those of other delegations, were reflected in Principle 6: "The just struggle of the peoples of all countries against pollution should be supported" (UNCHE 1972). In addition, Principle 26 states: "Man and his environment must be spared the effects of nuclear weapons and all other means of mass destruction. States must strive to reach prompt agreement, in the relevant international organs, on the elimination and complete destruction of such weapons" (UNCHE 1972).

Of course, many issues that currently fit into our cognitive environmental framework were also included in the declarations. Aspects of the preservation logic were echoed in Principle 2, which declared that "The natural resources of the earth including the air, water, land, flora and fauna and especially representative samples of natural ecosystems must be safeguarded for the benefit of present and future generations" (UNCHE 1972). The resource management perspective was also highlighted, as in Principle 5: "The non-renewable resources of the earth must be employed in such a way as to guard against the danger of their future exhaustion" (UNCHE 1972). And the pollution agenda was represented in Principle 6: "The discharge of toxic substances or other substances ... in such quantities of concentrations as to exceed the capacity of the environment to render them harmless, must be halted in order to ensure that serious or irreversible damage is not inflicted upon the ecosystem" (UNCHE 1972).

The concerns of the global South for economic development and growth were also prominently embodied by the environmental agenda in several principles. For instance, Principle 9 stated: "Environmental deficiencies generated by the conditions of under-development and natural disasters pose grave problems and can best be remedied by accelerated development through the transfer of substantial quantities of financial and technological assistance as a supplement to the domestic effort of the developing countries and such timely assistance as may be required" (UNCHE 1972). Principle 11 similarly proposed: "The environmental policies of all States should enhance and not adversely affect the present or future development potential of developing countries" (UNCHE 1972).

Accordingly, the 1972 UN Stockholm Conference provided an important workspace in which the cultural conceptions underlying the global environmental regime were developed. The initial set of documents did not provide a clearly worked out and logically consistent agenda for environmental protection. Instead, as hypothesized by the world society perspective, the conference and its documents suggested a first step in the development of a global environmental agenda and a modern conception of the environment.

The Global Environmental Institution

The creation of a new institutional structure for global environmental issues was a formidable achievement that would dramatically impact the effectiveness of environmental protection around the world. As Lynton Caldwell summarizes, "The primary accomplishment of the Stockholm Conference was the identification and legitimization of the biosphere as an object of national and international policy" (1984, 53). The creation of the United Nations Environment Programme signaled that "environmental problems of broad international significance" fell within the province and competence of the UN network" (McCormick 1995, 131). The construction of the ecosystemic framework for environmental protection created a broad umbrella linking diverse aspects of the environment and bringing a global perspective to issues formerly perceived as local. Consequently, this broad cognitive framework has enabled the development of an expansive global institution.

First, the 1972 UN Stockholm Conference created a global organizational structure with a mandate to deal with broad environmental issues. The formation of the United Nations Environment Programme, which was passed by the UN General Assembly in December of 1972, ushered in an explosion of growth in international environmental structures (J. Meyer et al. 1997b; Frank et al. 2000a). Many scholars have noted the flood of international treaties, national legislation and environmental agencies, the growth of Green parties, and the proliferation of more than 20,000 national environmental interest groups that have followed the founding of UNEP (McCormick 1995, xii-xiii).

Second, the formation of a global environmental regime encouraged the proliferation of a multitude of agents tasked with environmental protection at international, national, and local levels, an issue taken up in Chapter 4. In particular, the formation of the United Nations Environment Programme encouraged an explosion of environmental international non-governmental organizations (INGOs) and inter-governmental organizations (IGOs) (J. Meyer et al. 1997b; Frank et al. 2000a). In addition, environmental scientific groups, social movement organizations, and concerned citizens have proliferated worldwide.

Third, cultural meanings of the environment changed with the 1972 Stockholm Conference and its expansive formulation of the ecosystem. The ecosystemic framework flexibly allowed the addition of new environmental problems to the agenda, encouraging continual expansion. The framework facilitated the inclusion of a wide variety of participants, including developing countries, scientists, and social movements as well as the participation of the industrialized West. Moreover, the framework provided a unique niche within the United Nations system that allowed the establishment of the United Nations Environment Programme. Thus the construction of the ecosystemic perspective was a major achievement of the 1972 UN Stockholm Conference.

Did the formation of all these global and national structures have an effect on environmental outcomes? To answer this question, the following chapters theorize and explore the effects of this global environmental regime on environmental protection efforts.

3

Institutional Structure

Since the 1972 UN Stockholm Conference, there has been an explosion of institutional structures focused on environmental change, including a wide variety of international environmental treaties and national environmental laws. Governmental organizations have also been created to oversee these regulatory policies at the national and international levels. Around the world, over 160 countries have established a national ministry of the environment, and more than 90 countries have developed a comprehensive set of environmental regulatory policies (Ecolex 2010; Statesman's Yearbook 2010). The United Nations lists over 200 international treaties on the environment, and other sources list over 120 intergovernmental organizations that are involved with environmental issues (UNEP 2005; UIA 2010). The growth of these institutional structures is shown in Figure 3.1.

To what extent do these international environmental structures bring about substantive improvements in environmental problems? A great deal of scholarship has taken a skeptical stance, raising doubts about the effectiveness of policy structures that are influenced or circumvented by corporations interested in maintaining the status quo. National policies, it is argued, represent "mere talk" that rarely impose the sanctions or enforcement mechanisms to coerce polluters to change their ways. International policies similarly are sorely lacking in terms of enforcement, seldom imposing significant punitive sanctions for wrongdoers. Consequently, skeptics have charged that environmental politics are rhetorical devices that allow polluters to carry out business as usual (Lindstrom and Smith 2001; Susskind and Ozawa 1992; Yearley 1991; Rowland 1973; Vogel 1995).

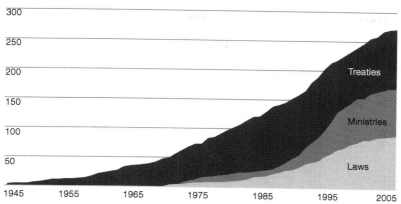

FIGURE 3.1. Institutional structure: Environmental treaties, ministries, and laws.

Such pessimism is generally based on studies that look for short-term, proximate effects of particular legislation or treaties, reflecting what Chapter 1 has labeled the Smoking Gun model of social change. Legislation is expected to have direct consequences for behaviors and the pace of environmental degradation. The numerous examples of unenforced, inefficient, or failed regulatory efforts fuel pessimism regarding the prospects for effective environmental reform (Lindstrom and Smith 2001; Susskind and Ozawa 1992; Yearley 1991; Rowland 1973; Vogel 1995).

This chapter develops a world society argument that focuses on indirect and long-term effects of national and international institutions. It is not disputed that regulatory structures typically fail to accomplish mandated requirements, as the literature repeatedly shows. If one surveys the big picture, however, evidence demonstrates that institutional structures are associated with dramatic real-world improvements over the longer term (Schofer and Hironaka 2005; Shandra et al. 2009; Shorette 2012; Cole and Ramirez 2013). These seemingly contradictory findings can be reconciled by turning to the Bee Swarm model of change described in Chapter 1. Institutional structures may appear ineffectual when their immediate consequences are scrutinized in the short term, but can nevertheless generate substantial macro-level social change via a host of different mechanisms.

How do institutional structures, such as international treaties or national laws, generate social change? In brief, institutions create a "workspace" for an environmental issue. First, institutional structures place the issue on the political agenda. When an environmental treaty is enacted, the solution for the environmental problem is often unknown. The creation of a treaty or law is the starting point that gives official recognition and attention to an issue and bootstraps subsequent efforts at research and the identification of potential solutions. Second, workspaces bring together relevant actors on an issue – governments, corporations, environmental activists, and even groups that oppose reform. Although counterintuitive, the inclusion of reluctant or opposing groups into the workspace often proves essential for the subsequent success of environmental reforms. Third, the long-term durability of institutional structures provides incentives to invest in technological or organizational innovations that might be costly in the short term. Taken together, these processes can generate social change, even if the particular treaty or law is weak, poorly conceived, and lacking in powerful sanctions. The world society perspective focuses on these broad consequences of institutional structures as long-term drivers of real-world improvement of environmental issues.

The chapter begins with an overview of the institutional structures that proliferated worldwide following the United Nations Stockholm Conference in 1972. Next, the chapter reviews the pessimistic prior research on the effectiveness of international treaties and domestic legislation. Studies have found that insufficient political will or managerial capacity often renders treaties and laws ineffective in achieving the intended goals. Treaties and laws do not appear to be the "smoking gun" that directly leads to pro-environmental change. The chapter next offers a world society analysis of institutional structures and their consequences, with emphasis on the concept of workspaces. This perspective does not contest the prior research that questions the effectiveness of treaties and laws. However, through the creation of a workspace and related indirect and longer-term processes, institutional structures may nevertheless generate significant improvements in environmental conditions.

The chapter concludes with an extended case example of international regulatory efforts to address ozone depletion. Global restrictions on the production of ozone-depleting substances are the result of one of the most successful global environmental regimes to date. However, the effectiveness of international efforts seemed dubious at the outset. The establishment of multiple international workspaces eventually bootstrapped the near-elimination of the production of chlorofluorocarbons and other ozone-depleting substances across the globe.

The Proliferation of Institutional Structures

The world society tradition has examined the expansion of institutional structure in various policy arenas. Scholars have observed the rapid propagation of international treaties, national laws, national constitutions, and governmental organizations in domains including human rights, the environment, education, and law (Frank et al. 2000a; Boyle and Preves 2000; Cole 2005; Frank 1997; Hafner-Burton and Tsutsui 2005; Frank 1999; Hironaka 2002; Hironaka and Schofer 2002; Wotipka and Tsutsui 2008). World society research has emphasized that these policy structures frequently involve top-down dynamics. Once a global regime is established, nation-states build national structures and conform to international norms or standards in any given domain.

World society research on the global environmental regime has observed this same pattern: the structuration of international institutions precipitated the diffusion of policies to nation-states around the world. While a handful of Western nation-states had nascent environmental regimes in the late 1960s, the majority of countries developed environmental policies in response to the formation of international environmental institutions (J. Meyer et al. 1997b; Frank et al. 2000a; Frank, Hironaka, and Schofer 2000b; Hironaka and Schofer 2002). This section briefly outlines three scholarly explanations for the proliferation of these international and national treaties and legislation on the environment.

Modernization Theory

Modernization arguments claim that regulatory structures are motivated by public concern over the quality of their living conditions (Inglehart 1977; Mol 1997; Rosa et al. 2010). Historically, industrialization occurred at the expense of the natural environment. As citizens achieve high levels of affluence, however, they begin to value clean air, clean water, and natural parks, even at the cost of slower economic growth (Inglehart 1977; Inglehart 1990). Citizens may also have been alarmed by the risks to health and well-being introduced by their advanced technological society (U. Beck 1999).

The primary prediction of modernization theory is that environmental protection policies will follow industrialization and affluence. This has not generally been borne out. While the earliest environmental regulations appeared in affluent societies, scores of less-developed countries, such as Indonesia and Thailand, have also enacted national environmental legislation, as have rapidly industrializing countries such as Brazil and South Korea (Taylor et al. 1993; Guimarães 1991; Khator 1991). Moreover, individual attitudes around the world increasingly favor environmental protection, even within developing nations and among poor segments of the population (Dunlap and York 2008; Dunlap 1995). Environmental regulatory structures have diffused across the globe, not only in wealthy societies. The process has certainly not been conflict-free, and in some cases economic concerns have trumped environmental ones (as discussed further in Chapter 5). Nor have environmental policies been equally effective in all nation-states, as discussed later in this chapter. Nevertheless, environmental regulatory structures have appeared in an unexpectedly wide variety of countries.

Social Movement Arguments

The social movement literature offers a second explanation for the growth of pro-environmental regulatory structures that focuses mainly on domestic processes. Scholars suggest that national legislation and treaty ratifications should arise in those countries with vocal and effective social movement groups (Fisher 2003; Giugni 2004, but see Broadbent 1998). According to this perspective, governments may be aware of the need for environmental legislation but unwilling to challenge powerful anti-environmental economic interests. Citizen protests and social movement activism are needed to push governments to confront corporate interests and enact environmental policies. Domestic citizen groups and social movement organizations are potential sources of pressure in this scenario. In addition, domestic social movement organizations may obtain assistance and greater leverage via transnational social movement groups or other states, in what has been labeled the "boomerang effect" (Keck and Sikkink 1998; Khagram, Riker, and Sikkink 2002; J. Smith 1997).

The social movement perspective suggests that pro-environmental mobilization should precede the creation of national and international policy structures. Again, this expectation does not closely conform to the evidence. Highly organized and effective social movements often follow, rather than precede, the formation of regulatory structures (Broadbent 1998; Giugni 2004; Longhofer and Schofer 2010). As discussed in Chapter 4, effective coordination of action is difficult in the absence of institutions. Institutional structures construct social problems, set the public agenda, and legitimate mobilization. Drawing individuals together into a broader coordinated effort requires the translation of local issues into a common language and shared definition of environmental problems. Without this, social movement activists tend to focus on specific and idiosyncratic local issues – the noise from a proposed airport or the damage from a particular oil spill, for example (Freudenburg and Gramling 1994). Institutional structures, and the workspaces they create, bring concerned actors together and provide the formal standards necessary to build public consensus on an environmental issue. While social movement groups may exist without these structures, they are hampered in their capacity to motivate significant social change.

Indeed, historical accounts show that social movement groups had little impact on the initial enactment of environmental policies in the 1960s and early 1970s. While a handful of countries had vocal environmental social movements including the United States, Canada, Japan, and the United Kingdom, such groups did not play a central role in legislative efforts. R. Andrews (2006, 209), a prominent environmental scholar notes, "It would be tempting to attribute [U.S. environmental legislation of the 1960s] to the rising influence of environmental advocacy groups, or of mass public concern for the environment, but in fact all these statutes except the 1970 Clean Air Act were enacted before those [social movement] forces had coalesced at the national level."

World Society Theory

World society scholars argue that national policies and organizations typically develop in response to international pressures (J. Meyer et al. 1997b; Frank et al. 2000b; Frank, Longhofer, and Schofer 2007; Longhofer and Schofer 2010). The emergence of a global environmental regime, described in the previous chapter, placed environmental issues on national agendas. States were invited to attend international conferences and to confer on treaties concerned with environmental protection. Prototypes of environmental policies were derived from peer states and propagated via international organizations. Environmental debates often percolated from international to national and local discourses. Even states that chose to sidestep environmental responsibilities were increasingly held accountable for environmental problems by international organizations and rapidly expanding pro-environmental national constituencies, legitimated and emboldened by international agendas and discourses.

The UN Stockholm Conference on the Human Environment in 1972 was the key international event that led to an explosion of international and domestic environmental regulatory structures (see Chapter 2). Treaties and intergovernmental organizations, in particular, grew rapidly in the period since the Stockholm Conference. The effect on nation-states was equally dramatic. National environmental policy efforts were rare before Stockholm, but commonplace after. Nearly every comprehensive national environmental law and every national environmental ministry was created either during the run-up to the Stockholm Conference or in the post-Stockholm era.

For the handful of treaties and laws that preceded the Stockholm Conference, it is only in retrospect that they have been characterized as "environmental." Rather than prescient forerunners of the contemporary environmental movement, these early treaties and laws were developed under conceptual frameworks that predated the modern understanding of environmental problems (see Chapter 2). The United States was one of the few countries to have developed environmental legislation before Stockholm. The comprehensive National Environmental Policy Act was passed in 1969, shortly after the Stockholm Conference was announced. However, a prior series of federal regulations in the 1960s included the Clean Air Act of 1963, the Motor Vehicle Air Pollution Control Act of 1965, the Water Pollution Control Act of 1966, the Air Quality Act of 1967, and the Clean Air Act of 1970 (R. Andrews 2006, 208). At the time, these laws were perceived as the expansion of federal control over urban planning and public health that built on institutional structures from the turn of the century. Policymakers understood these measures as primarily ensuring sanitation and human health. Only with the understandings developed since the Stockholm Conference of 1972 have these laws been reconceptualized as protection for the natural environment.

For most countries around the world, the 1972 UN Stockholm Conference and the global environmental regime that resulted were the primary motivators for the establishment of pro-environmental policies. One environmental scholar notes, "Both in anticipation of the Stockholm conference and in its aftermath, numerous governments and intergovernmental organizations established ministries and departments focused on meeting environmental challenges" (Schechter 2005, 37). In Japan, fourteen pieces of environmental legislation were passed by the legislature nicknamed the "Pollution Diet" in 1970 (Brenton 1994, 30; Broadbent 1998). Western European states also passed a flurry of environmental legislation, albeit at a slightly slower pace (Brenton 1994, 30). Environmental ministries or agencies were established in fourteen industrialized countries between 1970 and 1972 (Brenton 1994, 30). Even in the United States, the Stockholm Conference spurred substantial policy activity. By 1972, more than half of the fifty states in America had passed environmental legislation, and the United States was spending nearly 2 percent of its GNP on pollution control (Brenton 1994, 30).

Lynton Caldwell, an advisor to the U.S. government during the preparations for Stockholm, describes how the influence of the Stockholm Conference encouraged governments to follow the lead of the United States and Western Europe in environmental protection:

During the years immediately preceding the Stockholm Conference, the example of new environmental laws and agencies established in France, Sweden, the United Kingdom, and the United States (among other countries) and the request of the United Nations Preparatory Commission for status reports from all countries on environmental policy, made possession of an environmental policy a status symbol – evidence that a nation belonged among the more advanced or advancing states of the world and not among the backward nations. (Caldwell 1984, 43)

There is little disagreement in the scholarly community that the 1972 Stockholm Conference on the Human Environment ushered in a period of feverish growth of both international and national policy structures. However, some are doubtful that these structures have led to actual improvements of environmental conditions. As one scholar skeptically argues, "in terms of formal product it is difficult to view Stockholm as much more than a cosmetic event" (Brenton 1994, 50). The next section reviews the scholarship on the effectiveness of environmental treaties and policies.

The Apparent Ineffectiveness of Environmental Regulatory Structures

The rapid expansion of pro-environmental policies and laws is not disputed. However, scholars have voiced a great deal of skepticism about their effectiveness. Research across several fields has repeatedly shown that regulatory structures – treaties, legislation, or governmental organizations – very often fail to produce the mandated outcomes. Environmental issues are no exception. As Lidskog and Sundqvist (2002, 79) grumble, "Although today there are more than 170 negotiated conventions their implementations leave a lot to be desired." Even more dourly, Rowland (1973, 33) characterizes international environmental conventions as "activities which amounted ... to fighting a fire with a thermometer."

The scholarly debate about the effectiveness of regulatory structures is centered in the political science literature on compliance. Although a variety of factors have been hypothesized to affect compliance, two perspectives have dominated this debate. The first is the enforcement perspective, which explains failure as the result of lack of political will. The second is the managerial perspective, which views ineffectiveness as a consequence of resource shortages.

Political Will and Policy Effectiveness

One model, termed the enforcement perspective in the political science literature, theorizes policy failure as largely resulting from insufficient political will (Vogel and Kessler 1998; Keohane and Levy 1996; Susskind 1994). States may sign treaties or enact legislation in order to create the appearance of environmental concern, but without any intent to carry out the policies. As Susskind and Ozawa (1992, 147) summarize, "The signing of a convention ... allow[s] symbolic but empty promises to substitute for real improvements."

This lack of political will may manifest in a number of ways. In the first place, states may pass tepid legislation that does not seriously threaten entrenched economic interests. For the reversal of environmental degradation, significant changes in environmental behaviors are necessary. However, legislative promises are easiest to keep when they deviate only minimally from the status quo. Scholarship in this tradition has shown that governments tend to be conservative in their adoption of international agreements, preferring commitments they can easily keep (Raustiala and Victor 1998). If states can avoid real changes in environmental behaviors, treaties and regulatory structures may not improve environmental conditions.

Second, even if treaties or laws demand significant environmental change, states may put little effort into implementation or provide only the mildest of penalties for violators. Indeed, among environmental treaties, weak enforcement has been the rule rather than the exception. Hurrell and Kingsbury (1992, 28) write, "It is widely acknowledged that implementation and enforcement has been the weakest part of international environmental law and related regimes." Environmental treaties generally lack provisions for sanctions beyond "polite if vigorous disapprobation" (Hurrell and Kingsbury 1992, 22). In general, states themselves are responsible for the administration of their own environmental policies, with little oversight from international organizations or other third parties (Keohane, Haas, and Levy 1993; Ringquist and Kostadinova 2005, 89). Consequently, anecdotal evidence suggests that cheating on environmental obligations is widespread (DeSombre 2007, 27).

Finally, even if states seriously enforce environmental laws, sub-state actors may still have incentives to disobey. Corporations may calculate that even if the government vigorously enforces a law, only a small number of offenders can be prosecuted. Alternatively, corporations may comply with the letter of the law to avoid sanctions but may avoid costly changes that would actually improve environmental conditions. The adversarial relationship between corporations and government in the United States has promoted this letter-of-the-law approach, although scholars note it has been "highly inefficient" in terms of improving environmental conditions (Vogel and Kessler 1998, 30). By contrast, societies with more cooperative relations between industry and the state, such as Germany, Japan, and the United Kingdom, may be more likely to see improved environmental outcomes (Vogel and Kessler 1998, 30).

From the enforcement perspective, regulatory structures that conflict with major economic interests may be expected to have little effect on environmental outcomes. The fundamental lack of good faith effort by those whose behavior needs to be altered may seem an insurmountable obstacle to substantive change. In such cases, regulatory structures are likely to play only a symbolic role. Consequently, Keohane et al. (1993, 18) argue, "Environmental politics is replete with symbolic action, aimed at pacifying aroused publics and injured neighbors without imposing severe costs on domestic industrial or agricultural interests."

Resources and Policy Effectiveness

A second perspective is the managerial model, which explains the failure of regulatory structures as a consequence of insufficient resources (Chayes and Chayes 2001[1993]; Chayes and Chayes 1995; Victor, Raustiala, and Skolnikoff 1998). Governments may be reasonably willing to carry out their legal obligations yet be foiled by a host of obstacles. The most common problem is lack of economic resources, but challenges also include shortages of trained personnel, effective procedures, or bureaucratic capacity. In the short term there may be failures in the empirical improvement of conditions similar to that predicted by the enforcement model. In the longer term however, the development of resources may increase the effectiveness of regulatory structures. Herring and Bharucha (1998, 423) summarize: "A certain amount of noncompliance is not willful, but rooted in common bureaucratic pathologies: lethargy, conservatism, rigidity, and arrogance, inter alia. There may be no cure for bureaucracy."

The managerial perspective argues that treaty noncompliance is primarily due to an inability of states, many of which are quite poor, to comply with treaty obligations. As Chayes and Chayes (2001[1993],

266) note, "Quite apart from political will ... the construction of an effective domestic regulatory apparatus ... requires scientific and technical judgment, bureaucratic capability, and fiscal resources." Treaties may be set up without operating funds; legislation might be enacted without expanding relevant budgets; government agencies may lack the manpower to carry out their mandates. Shortages in resources and state capacity are particularly acute for states in the developing world (Jänicke 1996; Jänicke and Weidner 1995; Jänicke 1990). However, such obstacles to environmental reform are quite common in industrialized democracies as well (Chayes and Chayes 2001[1993]).

First, states may suffer from a straightforward lack of economic resources and organizational capability, particularly in developing countries. Environmental issues usually rank relatively low on priorities of many governments. For less developed countries, basic security, economic development, and government stability are likely to be given greater attention than are environmental concerns (Hurrell and Kingsbury 1992, 30; Blaikie and Simo 1998). Even in highly industrialized states, environmental budgets may be affected by shifts in political agenda or economic downturns. For instance, resources and governmental efforts in the United States were significantly constrained by the political agenda during the Reagan and Bush presidential regimes (R. Andrews 2006).

Second, resource shortfalls may simply be due to the tremendous amount of work that regulatory regimes require. Critics note that even a decade after the passage of the 1972 U.S. Clean Water Act, only half of U.S. freshwater lakes and rivers had achieved the mandated water quality (Vogel and Kessler 1998, 19). One reason was that Congress had passed eight major new environmental legislative bills between 1969 and 1972. As Vogel and Kessler (1998, 21) point out, "Trying to eliminate as many environmental hazards as possible, and acting in great haste, the legislators on Capitol Hill instructed the EPA to set standards for *all* major air pollutants (1970) and water pollutants (1972), to regulate all pesticides (1972), to control solid-waste disposal (1976) and to eliminate the toxic substances among thousands of industrial chemicals (1977)." Unsurprisingly, the Environmental Protection Agency (EPA) was unable to accomplish these tasks in short order. Unable to cope with this overload, the EPA could not establish an administrative framework for the implementation of the 1972 Clean Water Act until 1976. Similarly, the 1976 Resource Conservation and Recovery Act, which is discussed in Chapter 4, was not implemented for six years after its enactment (Mazmanian and Morell 1992). Considering the complexity of the issues involved, one might claim even these delayed achievements were quite a feat, even for America at the peak of its economic and scientific prowess. Other countries had even less in the way of resources and expertise to facilitate implementation.

Similar administrative overloads are evident in European countries and the international community. In the European community, more than 500 environmental directives had been legislated by 1998, with 100 new directives being added each year (Vogel and Kessler 1998, 22). Akin to the U.S. experience, many European environmental ministries were unable to sustain the pace of legislative implementation. In the international sphere, scholars fear that "international accords are being developed so rapidly that some analysts now raise the specters of "treaty congestion" and insufficient coordination among agreements (Victor et al. 1998, 1).

A third challenge is lack of organizational resources to oversee the multiple levels of actors involved in environmental behavior (Evans, Jacobson, and Putnam 1993; Victor et al. 1998). National governments accept responsibility for environmental behaviors by signing treaties and enacting legislation. However, the national government may be unable to guarantee the activities of subordinate actors such as corporations, city or state governments, or citizens who must carry out the regulations. This dilemma has been described as double-edged diplomacy: states face pressure to comply at the international level, yet they have limited control over the behaviors of subnational actors (Evans et al. 1993). As Victor et al. observe, "One reason implementation of international environmental commitments is so complex is that the paths to the targets' doors frequently weave through many intermediate actors and institutions" (1998, 4–5). Non-compliance may be the result of the complexity of coordinating international obligations with domestic institutions.

From the managerial perspective, states may have the political intent to carry out substantive environmental reform, yet lack resources on a variety of fronts to make those reforms. Despite this more optimistic stance, the managerial perspective generally agrees with the enforcement model on the bottom line, that the success record of pro-environmental regulatory efforts is poor. Rather than contest this claim, the world society perspective sidesteps this debate. As the next section outlines, the development of structure serves as a starting point for social change rather than an end point.

Institutional Structure and the Creation of Workspaces

This chapter argues that institutional structures have consequences well beyond their specific legal or regulatory requirements, regardless of whether these requirements have been effectively implemented. According to the Bee Swarm model, improvements in environmental conditions occur in an indirect and often halting fashion, as a variety of pressures are brought to bear on societies. Institutional structures, such as laws, treaties, and governmental organizations, provide channels for social activity by creating various workspaces addressing a given issue. The following section discusses three important dynamics: (1) agenda setting, (2) creation of workspaces for participants, and (3) the long-term effects of persistent institutional structures. These all serve to construct a particular social problem and mobilize activity and agents, which generate social change over the long term.

Agenda Setting

Institutional structures are central to the social construction of an issue in the public sphere by officially defining and publicly recognizing a social problem. As discussed in Chapter 2, social problems do not simply present themselves neatly delineated in objective reality. Instead, a social problem must be identified and labeled, its boundaries defined, and it must be placed on the political agenda in order to be recognized as a social problem. As Gusfield (1981, 3) wrote, "Human problems do not spring up, full-blown and announced, into the consciousness of bystanders." Policy structures cement social recognition of an environmental issue. Once on the political agenda, activity can begin to be coordinated over the issue.

Institutional structures facilitate agenda-setting in three ways. First, placing an issue on the political agenda allows scientific expertise and attention to be mobilized on a particular physical process (Young 1989a). The modernization perspective assumes science as the unproblematic foundation for environmental problems. However, directing scientific attention and translating scientific findings into language suitable for policymaking requires resources and effort that are more often the consequence of the policymaking process than the cause of it. More than one finding from a single team of researchers is needed to convince politicians to legislate potentially costly new regulations on a global scale. Studies, often funded by intergovernmental organizations (IGOs) or governmental ministries, may require the coordination of efforts of multiple scientific teams in multiple countries or universities. These large-scale scientific endeavors are usually necessary to support major policy innovations. Institutional structures are a prerequisite for motivating scientific research in a suitable policymaking format.

Second, official recognition motivates data collection on environmental degradation. Without the impetus of institutional structures, states rarely collect data on environmental problems, rendering them obscure and poorly understood. Data collection may require significant technical and organizational capability. Atmospheric pollution data, for instance, require measurement at multiple sites with accepted scientific methodologies organized by a centralized agency. These demands may be beyond the capabilities of many states. Developing countries, in particular, face "a severe dearth of the requisite scientific, technical, bureaucratic, and financial wherewithal" to comply with the reporting requirement of treaties "without technical assistance from the treaty organization" (Chayes and Chayes 2001[1993], 266). Even highly industrialized countries may not expend the required resources for data collection until mandated by treaty. For instance, it was nearly two decades before European states were able to develop an adequate system to collect the data on air pollution that was mandated by the Long Range Transboundary Air Pollution treaty (Raustiala and Victor 1998, 681).

Third, as previously discussed, the entrance of an environmental issue on the international agenda is likely to also put it on national agendas. State involvement in international conferences, scientific task forces, or treaty negotiations creates both capacity and interest in an issue at a domestic level. States involved in these conferences often sign treaties that further obligate national compliance. As Keohane et al. (1993, 15) observe that, "Western European states became more concerned about acid rain through the operation of [international] institutional activities, and that developed countries became more concerned about stratospheric ozone depletion due to activities connected to the Montreal Protocol."

Even states that choose not to participate in international conferences or refuse to sign environmental treaties may find an issue pushed onto their national agenda. Domestic environmental groups or transnational organizations put pressure on national governments that have sidestepped formal treaty obligations (Keck and Sikkink 1998; Khagram et al. 2002). For instance, the refusal of the United States to sign the Kyoto Protocol was seen by many as an end to efforts at greenhouse gas abatement in this country. However, as discussed in Chapter 5, the prominence of climate change on the international agenda has meant that the issue has been repeatedly brought forward on the domestic environmental agenda despite the lack of formal treaty obligations in the United States.

As Lukes (1974) has argued, the ability to place (or squash) issues on the agenda is one of the most important exercises of power. The setting of the agenda in the international arena is critical, in part because it is difficult for a domestic coalition to quash an issue that is prominent on the international stage. The recognition of an environmental issue on the international agenda is an important first step that opens up the possibility for action to take place, even if direct improvement in outcomes has not yet occurred.

The Creation of Workspaces

Institutional structures provide workspaces that bring relevant groups together to develop solutions for a given social problem. The enforcement model of compliance assumes that the solution to an environmental problem is already known and all that is needed is compliance with the solution. In contrast, the world society perspective suggests that institutional workspaces encourage the activity needed to construct problems and work out solutions. Even if an environmental problem is prominent on the agenda, the economic, technological, and political resources needed for finding solutions need to be marshaled. This section will examine three characteristics of the workspaces created by institutional structures.

First, workspaces bring together the ingredients – resources, agents, and organizational attention – for work directed toward the social problem. Laws and treaties are often passed without a clear understanding of how new regulations will be put into place. Frank Dobbin (2009) has argued that bureaucrats figure out the guts of a policy during implementation, rather than during the legislative struggle to pass a bill. New technologies may be needed that could not have been foreseen before legislative enactment. Rationalized administrative procedures may be developed, such as the invention of environmental impact statements or other bureaucratic protocols. Scientific studies may be needed to understand the highly complex processes involved in environmental damage and cleanup. Institutional structures establish workspaces in which these activities can occur.

Second, institutional workspaces bring reluctant or oppositional groups into dialogue with environmental proponents. In the absence of a workspace, reluctant groups can ignore an issue entirely or stall progress toward remedies. States with strong domestic contingents lobbying for the status quo are unlikely to change without external pressures.

Conversely, environmental proponents might enact stringent regulations domestically, but have little leverage to encourage laggard states to follow suit. The creation of a common workspace, in which environmental laggards participate in hopes of slowing down change, often serves the purpose of bringing laggards into the conversation.

Critics have often assumed that the inclusion of anti-environmental interests in institutional workspaces dilutes the resulting treaties or regulatory structures. This is problematic from the Smoking Gun perspective, which imagines that specific regulations determine subsequent environmental outcomes. The Bee Swarm argument suggests a different interpretation. The particular requirements or sanctions of a treaty may be less important than the general effects of setting global and national agendas, encouraging dialogue and motivating efforts to find solutions. While anti-environmental states might dilute a treaty protocol, the process of negotiation importantly nudges reluctant states toward addressing problems that would not happen otherwise. Instead, treaty laggards often face increasing pressure over time – from other states or social movement organizations – to improve their behavior. Paradoxically, a weak treaty that engenders widespread participation may have greater consequences than an optimal treaty that has few signatories.

Third, institutional workspaces, perhaps more than anything else, produce a great deal of talk and bureaucratic paperwork. Barnett and Finnemore (2004, 34) note that international organizations "tend to craft rational-legal solutions" and favor "[s]olutions that involve regulation, arbitration, and intervention by rational-legal authorities." The world society perspective suggests that these rational-legal solutions – although not as appealing to environmentalists as direct substantive requirements – may result in improved environmental outcomes in the long term. Risse and Sikkink, for instance, have identified a norm cycle in which environmental opponents initially adopt cosmetic changes to pacify criticism (1999, 25). However, in a dynamic environment with myriad pro-environmental agents, official environmental rhetoric may unleash further pro-environmental efforts. As Risse and Sikkink (1999, 27) argue, "Leaders of authoritarian states ... tend to believe that 'talk is cheap' and do not understand the degree to which they can become 'entrapped' by their own rhetoric.... By the time they realize their mistakes, they have already unleashed forces of opposition beyond the expectations of the regime, and the situation is often out of their control."

In sum, the creation of formal legislative structures – and short-term implementation – is a starting point rather than an end point. The creation of a workspace bootstraps numerous processes – agenda setting, construction of problems, and the search for solutions. It may appear that environmental efforts regress one step for every two steps moving forward, as states expend their efforts on bureaucratic solutions that have little impact on environmental outcomes. However, from the world society perspective, even these ineffective efforts create expectations and pressures on nations for improvement over the longer term.

Persistence of Institutional Structures

A third key feature of institutional structures is their durability, or persistence over time. In many cases the initial set of standards or rules are insufficiently implemented or are inadequate to address the magnitude of the social problem. However, even inadequate standards may create incentives for further change that would not otherwise have existed. New rules and standards may not directly force improved environmental conditions, but may nevertheless create formal and informal pressures for polluters to alter their behavior.

First, the existence of formal policies enables a wide variety of environmental agents to mobilize on behalf of improved environmental conditions. Once a law or treaty has been enacted, pro-environmental agents have enhanced legitimacy to encourage compliance. States are reluctant to sign treaties because official sanction can unleash a flood of public demands for compliance. Even if governments are reluctant to enforce environmental regulations against polluters, citizen groups can wield a law against the government or corporations once a law has been formally enacted (Hironaka and Schofer 2002). Such efforts may be transnational in scope as well. International organizations may invoke "shame and blame" tactics to encourage states to comply with treaty requirements (Victor et al. 1998). Transnational social movement groups may arouse broader public concern about environmental transgressions in other countries. Such efforts would be much less effective without the presence of accepted formal standards by which a state can be shown to be inadequate.

Second, institutional structures create the conditions for the development of more institutions. As discussed in Chapter 2, institutional structures tend to beget more structures. In the environmental arena, subsequent treaties with higher standards have often replaced a weak international treaty on the same issue (Brown Weiss 1998). National laws may be followed by further legislation that enlarges the scope of pro-environmental behavior. New organizational structures are created to implement or oversee the new environmental regulations, which then seek to extend their bureaucratic purview. As Levy, Keohane, and Haas conclude, "Institutions are needed early on to help create the conditions that make strong rules possible" (1993, 413).

Third, the promise of durability offered by institutional structures encourages the reconstruction of interests. Treaties and laws provide assurance that a regulatory regime will endure long enough to make short-term costs worthwhile. This promise of durability creates incentives for innovation. As is widely recognized, the implementation of environmental solutions may prove economically costly in the short-term. For instance, Levy et al. (1993, 413–414) note, "Many of the mechanisms that facilitate adoption of effective rules and meaningful national implementation of joint rules are time-consuming to create." Structures such as laws and treaties promise that environmental regulations will extend for the long-term, making it worthwhile to expend costly effort and resources in the short term. This promise of durability reshapes the interest of states and firms, as discussed in Chapter 5.

In sum, institutional structures devoted to environmental issues may not be a direct solution to environmental problems. Instead, structures can be read as a commitment to undertake work on behalf of the environmental issue. Implementation may be seen as a "perpetual cycle of policy" as new information, technologies, and political pressures are generated (Victor et al. 1998, 6). The value of the structure lies in its ability to provide workspaces that set the agenda and enable work on the development of solutions. These workspaces allow the coordination of action, bring together pro- and anti-environmental agents, and assure the durability of the institutional regime. The following section illustrates the importance of institutional workspaces, focusing on the example of the international ozone regime.

Institutional Structures, Workspaces, and the Global Ozone Regime

One of the major successes of the global environmental regime has been the regulation of ozone-depleting substances worldwide. Such a feat could not have been accomplished by a handful of states independently legislating their domestic production of chlorofluorocarbons (CFCs). Instead, a global regime was necessary to bring together initiator and laggard states, industry and environmental organizations, current consumers and future producers. As Edward Parson observed, "Because a dozen countries produced CFCs, and because both CFCs and ozone were mixed globally, even the largest producers could not solve the problem alone" (2003, 43–44).

The formation of this international institutional structure provides a clear illustration of world society processes. Modernization theory would expect that the scientific recognition of the problem of ozone depletion provided an objective basis for the motivation of international activity. However, no specific scientific breakthrough preceded the formation of the 1987 Montreal Protocol. Parson (2003, 249) writes, "At the time of the transition to regime formation, from late 1986 through 1987, there had been no scientific advance that significantly increased the confidence that CFCs had caused or would cause ozone loss." In contrast, the world society model argues the reverse: international structures and proposed policies were necessary to place the ozone problem on the international agenda and coordinate the scientific research to fully understand the issue.

Social movement theorists might emphasize the influential role of environmental groups, public opinion, and the media on policymaking. In the case of the ozone reduction regime, these factors did soften government resistance in certain countries, especially in the United Kingdom and West Germany. However, these social movement actions flourished within the structural framework of proposed international treaties rather than providing the original impetus. Moreover, social movements were not prominent in the majority of states that signed the Montreal Protocol. Conversely, the world society perspective highlights how the emergent treaty created a workspace in which a variety of participants were brought together. Social movement groups were among those involved, but were not more important or essential than other proponents of CFC regulation.

The world society perspective highlights the development of international structures around the issue of ozone depletion that created workspaces directed toward the eventual reduction of chlorofluorocarbon production. The following section outlines three major impacts of institutional structure. First, treaty negotiations set the agenda and supported scientific scrutiny and data collection that kept the ozone issue on national agendas. Second, international structures created conferences and other workspaces that assembled participants with varying interests and provided the time and space to develop solutions. Third, although initial standards for CFC reduction were weak, they paved the way for successive treaties that imposed much more demanding standards.

International efforts to address the ozone layer have been hailed as a success. Indeed through the lens of hindsight, the ozone regime may seem like a trivial case. However, the success of this international regime was by no means preordained. As one environmental scholar observed, "By the late 1990s, the achievements of the ozone regime were so remarkable that it was becoming easy to forget the years of effort that were required to establish the regime" (Parson 2003, 245). This section recounts the activities within institutional structures over two decades that eventually resulted in a worldwide ban on chlorofluorocarbons. The extensive secondary source material available for this case allows a close examination of the impact of institutional structures.

Setting the Agenda

The initial scientific discovery of the effects of chlorofluorocarbons on the planetary ozone layer occurred in the early 1970s. However, much work had to be done at national and international levels before laboratory findings could serve as the basis for a global institutional regime. Preliminary to formal treaty negotiations, institutional structures, such as international organizations and national states, underwrote the huge mobilization of scientific expertise and massive data collection effort that was necessary to study the issue and justify the creation of a regulatory regime. These efforts brought the issue of ozone depletion to the attention of many countries and widened the scope of the issue to include future as well as current producers of CFCs. In the case of the global ozone regime, a decade of agenda-setting occurred before formal international treaty negotiations were underway.

In the first place, scientific expertise had to be mobilized to provide the foundations for the global ozone regime. Modernization theorists such as Haas (1992) have argued that the scientific consensus on the effects of CFCs created an epistemic community that provided the basis for the development of policy structures. In contrast, the world society perspective points out that the resources and scientific attention needed for scientific consensus did not arise spontaneously. Instead, scientific findings on the ozone were the result of national and international structures that marshaled the resources and coordination necessary for the development of a scientific understanding of the problem.

In 1974, two scientific papers were published that raised alarms about the potential effects of chlorofluorocarbons on ozone molecules. One study by Molina and Rowland found that chlorofluorocarbons, an increasingly common industrial chemical, could be broken down by solar radiation to release large quantities of chlorine molecules (Stoel, Miller, and Milroy 1980). Another study by Stolarski and Cicerone found that a single chlorine atom could break apart thousands of ozone molecules, yet the authors could foresee no reason that large amounts of chlorine would be present in the atmosphere (Litfin 1994). When these two studies were put together, scientific alarm was raised about the potential effects of chlorofluorocarbons on the ozone layer of the earth's stratosphere.

However, it was a big step to translate laboratory findings to the scope of planetary atmospheric processes. The magnitude of research needed to establish the planetary consequences of ozone depletion could not have been carried out without the support of major intergovernmental organizations. Not only did the original findings need to be replicated, but the complex dynamics of the upper atmosphere needed to be taken into account. While chlorine atoms might destroy ozone molecules in a controlled lab setting, the theorized effects of chlorofluorocarbons might not have a measurable effect on the vast scale of the planetary atmosphere. Nor could the early studies account for potential interaction effects of the various other gases present in the atmosphere. Ozone molecules make up only a tiny proportion of atmospheric gases, and they fluctuate wildly by season and geographic location due to natural dynamics (Benedick 1998). Research on these topics could not be produced by one or two teams of scientists, but required hundreds of experts in multiple specialties.

Consequently, Benedick (1998, 5) observes, "The best scientists and the most advanced technological resources had to be brought together in a cooperative effort to build an international scientific consensus." The governments of the United States, the United Kingdom, West Germany, and the European community sponsored domestic scientific studies (Parson 2003; Benedick 1998; Litfin 1994). Intergovernmental organizations also funded studies, including one in 1984 that was cosponsored by the National Aeronautics and Space Administration (NASA), National Oceanic and Atmospheric Administration (NOAA), the Federal Aviation Administration (FAA), the United Nations Environment Programme (UNEP), the World Meteorological Organization (WMO), the West German Ministry for Research and Technology, and the Commission of the European Communities. This study was based on a yearlong effort of a taskforce composed of about 150 scientists worldwide (Benedick 1998, 14). Another study in 1986 was cosponsored with the U.S. EPA and resulted in "a weeklong international conference on risks to human health and the environment from ozone loss and climate change" (Benedick 1998, 20). These and other international studies were needed to provide the basis for a scientific epistemic community that convinced the international community of the validity of the original scientific findings.

Second, data was lacking on the amounts of chlorofluorocarbons being emitted into the atmosphere. Most countries did not require companies to report CFC emissions, nor were technologies initially available for measuring CFC emissions in the atmosphere. Corporations either held such information as a trade secret or simply did not keep records. Estimates of future CFC production were even more uncertain. However, models of potential ozone depletion required reasonable estimates. Without estimates and sensible projections of current and future CFC emissions, the size of the ozone depletion problem could not be ascertained.

Data collection did not occur until mandated by national legislation and international treaties. The United States, the world's largest CFC producer, required reporting of CFC production in its 1977 legislation. A few other countries such as Sweden, Norway, and Canada similarly required CFC reports according to national legislation. However, it was not until the Vienna Convention of 1985 that many countries began to require CFC production statistics (Litfin 1994, 76). Because the treaty proposed a benchmark based on 1986 production levels, signatory states scrambled to collect that information. The exacting modeling process of planetary atmosphere dynamics required worldwide data collection efforts that could only be coordinated by international organizations (Miller and Edwards 2001).

Third, the creation of an international agenda on ozone depletion placed the issue on the national agendas of many states that were small or insignificant producers of chlorofluorocarbons at the time. Small CFC producers such as Argentina, Brazil, Egypt, Kenya, and Venezuela would have had little reason to consider limits on CFC production for several decades. However, the Montreal Protocol of 1987 framed the ozone depletion problem as one of worldwide scope. Because the longevity of CFCs was a key aspect of the problem, the Protocol claimed that future producers needed to be taken into account as well as current producers. The Montreal Protocol sought to curb future production for states that had little or no CFC production in the 1980s. Such attempts did not pass uncontested – India and China vehemently objected and refused to sign the Montreal Protocol. However, the Montreal Protocol did succeed in putting ozone depletion on the agenda of many developing countries in the global South that would otherwise have had no reason to consider the issue. As M. Hoffman (2005, 2) notes, "Before 1987, the *global* ozone depletion problem only required the participation of between twenty-five and thirty, mostly Northern states in governance processes. By 1990, the international community understood the ozone depletion problem to require *universal* participation, and more than 100 states participated in governance activities through the 1990s."

By the mid-1980s, the issue of ozone depletion was squarely on the international agenda. Owing to the efforts of international organizations such as the OECD and the European Union, there was scientific consensus on the ozone-depleting effects of chlorofluorocarbons. However, scientific agreement did not translate into political agreement. Instead, there was considerable reluctance to develop international restrictions on CFC production on the part of major producers such as France, the United Kingdom, and West Germany.

Creating Workspace

The proposed and actual formation of international treaty structures created multiple workspaces around the issue of ozone depletion. These workspaces brought together pro-environmental and anti-environmental diplomats, industrial interests, environmental groups, and social movement protesters through the negotiation processes for the Vienna Convention and Montreal Protocol. During the late 1970s and early 1980s, little progress was apparent. An endless series of meetings appeared to produce little beyond reams of diplomatic discourse. Through these workspaces, however, reluctant states and industrial corporations were eventually brought on board the ozone agenda. Although the eventual result was the Vienna Convention of 1985, this treaty provided little more than additional talk. While rhetorically proclaiming the importance of ozone depletion, the Vienna Convention provided no requirements for enforceable standards or sanctions.

Although international discussion began in the mid-1970s on the potentially damaging effects of chlorofluorocarbons on the ozone layer, little progress in international negotiations was evident for the first decade. Parson (2003, 44) calls the first several years of international discussion an "unmitigated failure," arguing that pro-environmental states such as the United States, Sweden, Canada, and Norway "simply failed to mount enough pressure on the major CFC-producing nations to overcome their vigorous, organized resistance." By 1980, Stoel et al. (1980, 38) wrote of Japan, Australia, the Soviet Union, and Yugoslavia, "There is little indication that ozone depletion due to fluorocarbons has been recognized as a potential problem." Parson (2003, 110) observes that, "By early 1980, the initial campaigns for international controls of ozone-depleting chemicals had failed" and that "Even most of the environmental groups that had been active during the aerosol wars stopped paying attention" (2003, 60).

Nevertheless, treaty negotiations created several international workspaces that provided for the continuation of discussion despite the entrenched resistance of major CFC producers such as the United Kingdom, France, Russia, Japan, and the European community, and the seeming apathy of the rest. A UNEP conference in 1977 created a World Plan of Action on the Ozone Layer and also established the Coordinating Committee on the Ozone Layer (CCOL) (Brunnée 1988, 22). In 1981, UNEP sponsored multilateral negotiations on the issue of ozone depletion, followed in 1982 by "the Ad Hoc Group of Legal and Technical Experts for the Protection of the Ozone Layer" (M. Hoffman 2005, 11). Similar efforts were made in the European community and the OECD. The initial rounds of these negotiations were attended by only a handful of states. As Parson notes, "Conditions were not promising for starting international negotiations" (2003, 114).

Second, the promise (or threat) of international negotiations encouraged many groups with varying interests to participate in these international workspaces. Environmentalists are often skeptical about the presence of industrial lobbyists or laggard states in workspaces, fearing the dilution of the negotiations. Yet in the end, bringing together multiple interests may be even more important than the actual clauses of the resulting treaty. In the case of the ozone depletion regime, the threat of an international treaty brought laggards such as the United Kingdom and France into dialogue with pro-regulation states such as the United States and West Germany. Ignoring these anti-environmental interests in the hopes of creating a stronger treaty would have mistakenly assumed that the effectiveness of the treaty lay in the document rather than in the process.

In the ozone negotiations, the United States was a leader, having passed national controls on chlorofluorocarbons in 1977. Its influence was magnified because of its prominence as the world's largest producer of chlorofluorocarbons. Other states such as Canada, Austria, West Germany, and Norway eagerly hopped on the anti-CFC bandwagon, perhaps aided by their low production levels (M. Hoffman 2005). However, there was strong resistance from major CFC producers such as France and the United Kingdom. Responding to domestic industrial interests, France and the United Kingdom denied the need for regulation of chlorofluorocarbons and urged that European Union proposals provide symbolic declamations rather than require measurable decreases in output (Benedick 1998). Although these states participated in the international conferences, they were strongly resistant to a treaty that went beyond platitudes of environmental concern.

Third, although these conferences and workshops appeared to have little effect on substantive environmental outcomes, they contributed to the slow accretion of knowledge, structure, and the recruitment of environmental agents. As Parson (2003, 254) argues,

Even this modest increase in institutional structure granted continuity of proceedings, allowed the possibility of intersessional work to prepare negotiating agendas, and created an expectation that major nations would participate, present positions, be prepared to defend them substantively, and at least pretend to work toward agreement in good faith.... This increasingly dense web of meetings, requirements, and expectations maintained slight forward pressure at all times, and provided opportunities for large leaps forward when other conditions became favorably aligned, as they did in late 1986.

By the mid-1980s, the issue of ozone depletion had been the subject of a wide variety of international conferences, workshops, and meetings for nearly a decade. Yet all of this talk amounted merely to sound and fury. Progress on an international ozone regime appeared deadlocked. Eventually, the deadlock was broken by the slow buildup of structures at the international level – international conferences, committee reports, data collection, and workshops. Although these actions appeared merely symbolic at the time, these seemingly pointless activities provided the workspace for substantive technological, economic, and political change to occur, leading to a massive decrease in worldwide chlorofluorocarbon production.

The initial step was the ratification of the Vienna Convention of 1985, which was the first international treaty on ozone depletion. The Vienna Convention itself was widely viewed as an ineffectual treaty that was high on rhetorical promises but low on substance. The Vienna Convention provided no controls for CFC production, merely "establishing a general responsibility of states to protect the ozone layer" and "calls for various forms of scientific and technical cooperation among its parties" (Litfin 1994, 75). Nor was any particular substance specifically identified as contributing to ozone depletion (Benedick 1998, 45). By 1986 only six states had ratified the Vienna Convention – Canada, Finland, Norway, Sweden, the United States, and the Soviet Union – and these states already had stringent domestic restrictions on CFCs.

Parson gloomily concludes that the Vienna Convention was an "exhausted compromise," which "offered ... little concrete benefit"

(Parson 2003, 122). Litfin similarly agrees that the "greatest significance of the Vienna Convention was that it represented the first global consensus that there was indeed a problem" (1994, 77). However, the Vienna Convention did provide a launching point for a series of workshops that in turn established a framework for the Montreal Protocol of 1987 (Benedick 1998, 47).

Institutional Persistence

The third influence of institutional structure is the effect of institutional persistence on bringing about change. The Montreal Protocol of 1987 was acclaimed as an "unprecedented breakthrough" (Andersen and Sarma 2002, 93–94). As Parson (2003, 27) claims, "The ozone treaty is widely cited as the most successful example of international environmental cooperation to date and the best model for progress on such issues as climate change." However, such an achievement did not come about easily. The existence of institutional structure, in the form of treaty negotiations, allowed the additional voices of social movements and public opinion to demand improvements in line with the expectations raised by the treaty. The Montreal Protocol has subsequently been expanded and elaborated, so that further versions build stepwise on the advances of earlier versions. Finally, the promise of durability offered by the Montreal Protocol altered corporate calculations of interests, encouraging technological innovation. Although treaty negotiations had begun with few technological substitutes available for the many industrial uses of chlorofluorocarbons, the durability of the structure encouraged corporations to invest in research and development of alternatives.

First, the durability of institutional structures allowed public opinion to weigh in during the final stages of negotiation of the Montreal Protocol. Scholars have noted that public opinion waxes and wanes throughout the policy cycle (A. Downs 1972). The durability of institutional structures allows the mobilization of public opinion at critical moments, rather than requiring continuous public support throughout a decade of negotiation. The effectiveness of the public campaigns resulted from their ability to work within the process of international structural development. Social movement groups were particularly important in the United Kingdom for moving the government of Margaret Thatcher toward a more benign view of the Montreal Protocol. Scholars have argued that public opinion and social movements were also critical in West Germany, Canada, and the United States (Parson 2003, 36; Rowlands 1995, 229–230). As Benedick (1998, 5) notes, "The media, particularly press and television, played a vital role in bringing the issue before the public and thereby stimulating public interest," which "helped influence several countries to change their initial positions on the need for regulation."

Surprisingly, all of the world's major chlorofluorocarbon producers signed the Montreal Protocol in the end. Given the strong initial resistance by states such as the United Kingdom and France, such agreement was not foreordained. Although several factors influenced the decision to sign, one key factor was the role of public opinion. Although states such as the United Kingdom had been reluctant to agree to an international ban on chlorofluorocarbon production, the presence of the issue on the international agenda expanded the effectiveness of social movements, public opinion, and the media (Jachtenfuchs 1990, 275). Public opinion does not provide the whole story, since many of the signatory nations, including Japan, Russia, and France, did not have significant amounts of social movement activity on the ozone issue, yet signed the Montreal Protocol despite initial governmental reluctance. However, the prominence of the ozone issue on the international stage forced reluctant states to deal with the issue on their national agendas as well.

Second, the issue of ozone depletion illustrates the accretion of policy structures at the international level. The Montreal Protocol of 1987 was the first international treaty to set standards. It called first for a 20 percent reduction in chlorofluorocarbon levels beginning in 1993 for industrialized states (from a 1986 baseline), followed by a 50 percent reduction beginning in 1998[1] (Andersen and Sarma 2002, 85). This standard was the result of the intersection of scientific arguments, diplomatic negotiations, and industrial interests. Clauses subjected various sets of states to modified standards.

Although the creation of standards was deemed a success in itself, many scientists feared that even the substantial 50 percent decreases in CFC production mandated by the Montreal Protocol would be insufficient to address the problem of ozone depletion. Scholars have argued that the true success of the Montreal Protocol lay in the provision of institutional structure that allowed for the further expansion of policy structures. Signatories met again in London in 1990, Copenhagen in 1992, Vienna in 1995, back in Montreal in 1997, and Beijing in 1999 to expand the list of ozone-depleting substances and pledge to reduce

---

[1] For a set of developing countries, an additional 10% production in CFCs was allowed for exports during a ten-year transition period toward reduced CFC production. (Andersen and Sarma 2002, 85).

production levels even more. The progress achieved in these amendments to the Montreal Protocol was astounding. As Litfin notes, "Chemicals that only a few years before had been considered irreplaceable were now targeted for elimination" (1994, 11).

Third, the institutional durability offered by the Montreal Protocol was an essential component in the recalculation of economic incentives favoring the development of technological alternatives to chlorofluorocarbons. In the early 1980s, industrial lobbyists on both sides of the Atlantic denounced the astronomical estimated costs that would be imposed by a ban on chlorofluorocarbons, and the Montreal Protocol was adopted in 1987 "against the strongly expressed preferences of major industry actors in both the United States and Europe" (Parson 2003, 249). Even as late as March 1986, the chemical giant DuPont, one of the world's largest producers of chlorofluorocarbons, had asserted that there were "no foreseeable alternatives available" (Litfin 1994, 94).

However, once the Montreal Protocol had been signed, DuPont announced within a year that alternatives could be developed that would be only slightly more expensive. So abrupt was the industry about-face that cynics suspected that DuPont had been nursing alternative technologies all along. However, scholars have shown that the shift was not due to a new technological breakthrough but to new economic incentives brought about by the implementation of the treaty (Litfin 1994, 94). Once it became clear that an international treaty regime banning chlorofluorocarbon production would be implemented, corporate calculations shifted (a process discussed further in Chapter 5). DuPont later admitted that it had shelved research on CFC substitutes in the early 1980s because of the cost it would take to develop these alternatives (Parson 2003, 249). Calculations of the cost depended on assumptions of the availability of chlorofluorocarbons as a competitive alternative. If CFCs were to be phased out worldwide, market advantages would accrue to early leaders who could shift to non-CFC production as quickly as possible (Parson 2003, 249).

Consequently, the shift away from chlorofluorocarbons has turned out to be easier than initially expected. Technological solutions are not always so simple. Scholars have argued that chlorofluorocarbon regulation has been aided by the relatively small number of producers, primarily in highly industrialized countries, as well as the availability of alternatives. Other environmental problems, such as the regulation of greenhouse gases, may be less amenable (see Chapter 5). However, even if the regulation of chlorofluorocarbons has turned out to be relatively successful in hindsight, international treaty structures were essential to this success. Without the workspaces afforded by international treaty negotiations, it is doubtful that reductions in ozone-depleting substances could have been accomplished.

Conclusion

As one scholar has observed, even if only a few cars on a highway observe the legal speed limit, that does not mean that cars are behaving as if there is no speed limit (DeSombre 2007, 27). The development of institutional structures – laws, treaties, and government organizations – targeted at a particular environmental problem is an important step in promoting social change. These structures often do not have the direct effect on environmental outcomes posited by the Smoking Gun model. Instead, institutional structures provide the agenda, workspace, and promise of durability in which environmental solutions can be worked out. However, institutional structures are not necessarily sufficient in themselves. As the following chapters argue, institutional structures help create, legitimate, and institutionalize pro-environmental agents and cultural meanings that are also essential in promoting social change.

4

Agents

This chapter explores the nature of social actors and agency within world society theory in order to better understand the process of global environmental change. World society scholars have sought to sharply distinguish their structural approach from actor-centric theories in the social sciences (J. Meyer and Jepperson 2000). One consequence of this structural orientation has been the tendency for world society researchers to gloss over the role of particular organizations and individuals in the process of social change (but see Hallett and Ventresca 2006; Dobbin 2009). This chapter develops the concept of the "agent," in contrast to the conventional conception of an actor, to better characterize the activity that occurs within the context of institutional structures.

The Bee Swarm explanation of social change emphasizes the cumulative impact of many small influences rather than the decisive impact of particular regulatory structures or charismatic actors. Institutional structures create and empower an army of agents who implement, elaborate, and expand institutional structures over time. The aggregate effort of many agents working on behalf of the environment can add up to dramatic change in the long run. Agents play an essential role in social change, working under the aegis of existing institutional structures rather than cajoling from the outside.

Conventional arguments tend to reify institutional structures as fixed, monolithic, and impervious to change. Under the assumptions of the Smoking Gun model, fixed social structures are viewed as potent and stable mechanisms of social action. Given the strength of institutionalized capitalism in the contemporary world, the reorientation of society around environmental protection seems impossible. If structures are rigid, the hapless agent becomes a mere cog in the machinery, helpless to do anything but contribute to the momentum of the social structure. From this perspective, a Herculean actor is conceptually necessary to alter social structures and generate social change.

By contrast, the world society conception of institutional structure described in Chapter 3 fits with a correspondingly weaker version of the actor, here termed the *agent*. Institutional structures are skeletal and permeable frameworks. Rather than fixed on an inexorable trajectory, institutions are malleable to the efforts of agents, who elaborate and expand beyond their original premises. These changes and extensions often have a haphazard quality, as if renovations were being undertaken by hundreds of building contractors at once, rather than according to a single blueprint. The malleability of these institutional structures allows social change to occur routinely, and agents are central to the process.

This chapter develops the world society concept of agents by drawing a contrast with the conventional term actor. Here, "actors" represent an ideal type of individual or organization motivated by intrinsic interests and values. In contrast, "agents" enact roles enabled by institutional structures. Institutions imbue a wide variety of agents with purposes, identities, and goals (J. Meyer and Jepperson 2000). These agents are not merely automatons mindlessly carrying out directives from above. Agents actively interpret, elaborate, and expand institutional rules and discourses, sometimes generating new formulations and new fields. Working within the boundaries of a particular institutional context, agents engage in innovation and entrepreneurship. However, the efficacy of agents is inextricably linked to the institutional structures in which they operate. Agents draw on legitimated purposes and identities based on institutional regimes, as well as on organizational and material resources. These cognitive and material institutional resources have obvious implications for the effectiveness of agents, especially compared to "actors" whose agendas are internally generated and may not be consonant with existing social structures.

The following section elaborates on the theoretical distinction between actors and agents, and the corresponding image of agency that they imply. Second, the chapter examines how institutional structures create and empower a wide variety of agents. The chapter focuses in particular on international pro-environmental associations (INGOs), which have been central to world society research. A third section develops the world society conception of agency by exploring the ways agents contribute to social change.

Finally, the chapter turns to the empirical case of hazardous waste in the 1980s and 1990s to explore the concept of agents versus actors. Actors working outside the umbrella of early environmental and health institutions – exemplified by the so-called NIMBYs of the 1980s – were visible in the public sphere and were often characterized as a galvanizing force for change. This chapter offers a fresh perspective, arguing that NIMBY movements had relatively little impact on the broad issue of hazardous waste production and management. Instead a variety of institutional agents, including government bureaucrats, operators of waste management facilities, and citizen consumers, played a more important role in improving the handling of hazardous waste that has occurred over the past several decades. Indeed, historical evidence suggests that the NIMBY movements may have slowed the emergence of systematic measures to address hazardous waste.

Theorizing Actors and Agents

The world society perspective begins with the assumption that actors are socially constructed rather than endowed with *a priori* identities and interests (J. Meyer and Jepperson 2000; J. Meyer et al. 1997a; J. Meyer 2010; J. Meyer 1980; Suchman and Eyre 1992; March and Olsen 1984; Frank and Meyer 2002). Indeed, J. Meyer and Jepperson (2000) point out that the very conception of the "actor," and the propensity for individuals and organizations to describe and comport themselves as actors, is itself a cultural construction of the modern era. The term "agent" proves useful in highlighting these highly embedded entities, in contrast to the image of the actor with internally generated interests and goals. Agents are deeply embedded within institutional structures that provide a common cognitive and organizational framework. One might draw on the image of a craftsman artistically manufacturing a table. The craftsman puts skill and initiative into products and the end product may have a distinctive flavor, even as it reflects societal conventions or blueprints.

The world society perspective draws on the classic sociological insights of Cooley and Mead, in which individual identity is rooted in the "looking-glass self" (Cooley 1998; Mead 1934). The identity and interests of an individual are formed as a reflection of how she or he is viewed by others. Erving Goffman extended this perspective, pointing out that socially skillful individuals tend to match their behavior to the situation rather than acting out a prior self unconstrained by situational cues (Goffman 1974, 1959). Taken seriously, these arguments suggest that values, personality attributes, or strategic interests are not intrinsic to an individual. Rather, identity components are largely drawn from society, although agents may exhibit some flexibility and discretion in how they are deployed (Swidler 1986).

World society theory broadens this approach to explain the behavior of states and organizations, as well as individuals. International environmental treaties and regulatory structures, for instance, provide identities, meanings, and blueprints for action that shape behavior on a global scale. Institutional structures legitimate and empower a wide variety of pro-environmental agents that play an important role in social change.

Institutional structures provide solutions to several problems of social action. First, institutional structures construct purposes and interests for agents through the generation of pro-environmental identities (Snow et al. 1986; Snow and Benford 1988). Swidler (1986) argues that individuals draw upon a "tool kit" of available identities, roles, and strategies to organize their behavior. Frank and Meyer (2002) expand this argument, positing that the contents of the toolbox are drawn from the broader structures of society. Individuals may have always cared about the environment, and promoted it in various ways. However, without broader meanings and structures to draw upon, each individual was forced to reinvent his or her own role in environmental protection. The emergence of pro-environmental institutions effectively puts the cultural identity of "environmentalist," as well as various scripts or recipes for action, into the tool kit, making it easier for individuals to act accordingly. The dramatic proliferation of environmental agents in all sectors of society attests to the eagerness with which the environmental identity has been embraced. Where John Muir once stood alone, millions of individuals worldwide contribute to organizations such as the Sierra Club and the World Wide Fund for Nature today.

Second, agents draw upon institutional resources to build organizational structures and coordinate action (Zald and McCarthy 1980; Tilly 1978; McCarthy and Zald 1977; Oberschall 1973). Social movement scholars have found that social movement groups are more effective when they are able to build lasting infrastructure, such as networks of leaders, community centers, and a resource base (Gamson 1990; K. Andrews and Edwards 2005). This is much harder to accomplish in the absence of institutional structures, decreasing the effectiveness of movements (Piven and Cloward 1977). Institutions provide organizational structures and cultural models that facilitate movements and make social change easier. One might imagine efforts at environmental protection to be like a rowboat with twelve people at the oars. If each person paddles idiosyn-cratically, the boat is unlikely to go anywhere. However, when the oars are coordinated by a common organizational structure and set of cultural models and identities, the rowboat makes progress. Thus the world soci-ety perspective posits that effective action tends to occur within the con-text of institutional structures rather than from the outside.

Third, the magnitude of environmental problems requires the efforts of a wide range of agents to address. While citizens might improve local environmental conditions, many environment problems necessar-ily involve the cooperation of individuals, firms, and states in numerous countries to bring about substantive change. As Gould, Schnaiberg, and Weinberg (1996, 164) pessimistically conclude, "Locally based environ-mental movements may increase the public's awareness of environmen-tal issues, and they may create some change within local government and industry. But citizen-workers can protect their local environmental systems to only a very limited extent." Social movement scholars have recognized that broader political and social institutions are necessary to bring about macro-level change (D. Meyer and Staggenborg 1996). Given the difficulties of action without structure, social movement actors themselves are often pessimistic about the likelihood of substantive social change. Such pessimism supports the world society intuition that efforts toward social change are far more effective when they occur within exist-ing institutional structures.

The Creation of Institutional Agents

The global environmental regime legitimates the formation of an army of agents working to address a growing range of environmental issues. Environmental identities and scripts have proliferated in economic, political, and social domains, allowing individuals and organizations to attempt to be "green." This section describes a few of the wide variety of agents that have sprung up within environmental institutional structures, focusing particularly on international non-governmental organizations (INGOs). INGOs, which include a variety of voluntary associations, pro-fessional groups, and advocacy organizations, are a kind of agent that has garnered much attention by world society scholars (Boli and Thomas 1999). INGOs are often seen as the organizational embodiment of world society itself, and INGO membership is routinely used to measure the influence of world society on states as an indicator of the pressures of the Bee Swarm. This chapter explores the broad array of environmental agents that have been created, including direct and indirect ways that INGOs are agents of social change.

Environmental Agents

Along with the growth of international environmental laws, treaties, and governmental organizations there has been a dramatic proliferation of environmental agents. Today anybody can claim to be "eco-friendly" and take up a set of environmental identities and causes that would have been culturally meaningless a half-century ago. The abundance of these identities has been made possible by the development of institutional structures.

Environmental actors certainly predate existing environmental regulatory structures. Charismatic individuals like John Muir or Rachel Carson sought to address environmental issues long before the emergence of the global environmental regime. Environmental social movements in a few Western countries, including the United Kingdom, Australia, and the United States, preceded the modern environmental regime by a century or more (Hutton and Connors 1999; Weiner 2006). Likewise, pro-environmental NGOs predated the international regime in several industrialized countries (Johnson and McCarthy 2005; Frank et al. 2007; Bramble and Porter 1992). However in general, the growth of pro-environmental identities and organizations was hampered in the absence of global environmental institutions.

With the development of the global environmental regime, a wide variety of environmental agents proliferated (Frank et al. 2000a; Longhofer and Schofer 2010). Regulatory institutions create new formal roles and jobs for a host of pro-environmental agents. The environmental sector is a growing source of jobs with a projected 31 percent increase in this decade (CNN 2010). For these agents, addressing environmental problems – or at least generating more paperwork about the environment – is literally their job. Environmental laws and treaties create jobs for bureaucrats, lawyers, and engineers. Environmental ministries and government agencies hire civil servants to carry out their environmental oversight and regulations. Corporations need their own environmental experts to cope with the onslaught of new environmental regulations. Environmental programs have also expanded exponentially in universities, legitimating new professional expertise over environmental concerns (Frank et al. 2011; Bromley, Meyer, and Ramirez 2011).

The development of the environmental regime and the growing cultural salience of environmental concerns have also created economic environmental agents. Corporations in a variety of economic sectors have begun to produce environmentally friendly goods and services. This vibrant and expanding market for "green" products has been encouraged by consumers that have become increasingly willing to pay more for products marketed as environmentally friendly (A. Hoffman 2001; Yearley 1991; Shorette 2012). Environmental consultants confer as corporations seek to enhance the ecological substance, or at least the marketing image, of environmental values. Businesses are slowly beginning to view environmental regulations as creating the potential for profit (A. Hoffman 2001; Kamieniecki 2006). These activities can all be viewed as due to the efforts of environmental agents, whether or not an environmentalist identity is self-consciously adopted.

Environmental issues have also taken off in the political sphere. While many Western politicians viewed the early environmental movement of the 1960s with suspicion, every U.S. presidential candidate today claims to be pro-environmental, although specific policy approaches differ substantially across the political spectrum. Western European countries have had decades of influence by Green Parties (Doherty 2002; Kitschelt 1993; O'Neill 1997; Burchell 2002; Rootes 2003b). Even in countries without a formal Green Party, such as the United States, politicians are forced to confront environmental issues, since voters in many electorates place environmental concerns as one of their top political issues (Dunlap and Mertig 1992; Jacques, Dunlap, and Freeman 2008).

Society has also become broadly suffused with environmental concerns. Environmental curricula are increasingly mandated from elementary grades to college courses (Frank et al. 2011; Norgaard 2011). Media attention to environmental issues has grown exponentially over the past several decades, along with growing public interest (Andersen and Liefferink 1997). Celebrities champion eco-friendly products and practices. Individuals have a wide variety of actions they may utilize to express an environmental identity, ranging from reusing grocery bags and writing a check to an environmental organization, to living "off the grid" in a solar-powered home.

Finally, the expansion of environmental structure has encouraged the growth of environmental social movements (Hutton and Connors 1999; Rootes 2003a; Doherty 2002). Tsutsui and Shin (2008), focusing on human rights organizations, observe that development of global regimes encouraged the growth of local social movements worldwide. Longhofer and Schofer (2010) find a similar pattern among pro-environmental organizations: the global environmental regime has led to the proliferation of pro-environmental associations across the globe. Ignatow (2007) argues that preexisting ethnic, linguistic, and cultural groups provide the structural basis for environmental social movements in many countries. And recruitment and mobilization has been facilitated by greater public concern for environmental issues (Longhofer and Schofer 2010). Environmental social movements also find it easier to gain allies and media attention as environmental institutional structures develop and expand (Schaefer-Caniglia 2001). Thus in all but a handful of countries, substantial environmental movements developed after the initial creation of environmental structures in the early 1970s (Frank et al. 2000b; Longhofer and Schofer 2010).

In short, global environmental institutions have supported and empowered a wide variety of agents working on behalf of the environment. Along the lines of the Bee Swarm model, these environmental agents may not all contribute significantly to the solution of environmental problems. Some pro-environmental agents may have good intentions, but lack the resources or organizational capacity to substantively benefit the environment, as discussed in Chapter 3. Others may adopt pro-environmental stances but with little follow-through. Moreover, expansion of the global environmental regime also galvanizes opposition – anti-environmental agents, rooted in alternative institutional regimes. These issues of opposing interests and institutional conflict are taken up in Chapter 5. Nevertheless, a vast army of pro-environmental agents in a broad array of fields has indisputably emerged since the establishment of the global environmental regime in 1972.

INGOs as Agents

World society scholars have given particular attention to one type of agent: international non-governmental organizations (INGOs) (Boli and Thomas 1999; Boli and Thomas 1997). Empirical studies have traced the expansion of INGOs in a wide variety of domains and their effects on a daunting array of social phenomena. INGOs are not the only important agents in society. Multitudes of environmental agents are evident in various sectors, as previously discussed. Rather, the importance of INGOs lies in their capacity to bridge the gap between international and national activity, particularly in the global South. Pragmatically, INGOs have garnered attention simply because there is a good deal of information about them. By contrast, little systematic data on employees of environmental ministries, environmental engineers, environmental consulting firms, and other environmental agents are available. This section examines both the direct and indirect consequences of INGOs. Directly, INGOs affect environmental protection as agents in three spheres: civil society, the state, and the international community. Indirectly, INGO activity signals the extent to which international logics have penetrated a national society and serves as a proxy for myriad unmeasured activities of pro-environmental agents.

Wapner (2000, 89) defines non-governmental organizations as "political organizations that arise and operate outside the formal offices of the state, and are devoted to addressing public issues." These organizations are voluntary groups that are often composed of concerned citizens, rather than governments, corporations, or other officials. International non-governmental organizations (INGOs) are simply NGOs that have members in multiple countries (UIA 2010). Prominent environmental INGOs include Greenpeace, World Wide Fund for Nature, Friends of the Earth, and the Sierra Club.

For those living in North America or Western Europe, the activity of international non-governmental organizations (INGOs) may pass unnoticed, given the bustling commotion of domestic environmental agents. However, the abundance of domestic environmental agents in Western countries should not be taken as the rule throughout the world. Most states in the global South have a much sparser civil society landscape, with relatively fewer domestic organizations focusing on environmental promotion (Longhofer and Schofer 2010). In these societies, as well as in the international realm generally, INGOs are environmental agents of central importance.

INGOs play a central role in civil society, especially in non-Western states with weak governmental structures and impoverished civil societies (Bryant 2005; J. Brown 2006; Arts 1998; Betsill 2008; Wapner 1996; Bramble and Porter 1992; Longhofer and Schofer 2010). In non-democratic societies, INGOs often take the place of locally organized citizen groups. In some developing countries, for instance, environmental INGOs have pressured government officials on behalf of environmental protection policies (Wapner 2000, 94; Keck and Sikkink 1998). Environmental INGOs, with their links to the material and cultural resources of the international community, tend to be highly effective agents in developing country contexts.

In addition, INGOs undertake activities in developing countries that are usually expected of the government. In societies that lack government capacity, INGOs may be called in to draft laws on environmental protection or write position papers for international negotiations. As one scholar notes, in non-Western countries environmental INGOs and NGOS "take part in hearings, work in parliamentary commissions, and comment on bills. They participate at the municipal and state level in planning building projects and facilities. They have standing in court or support citizens in their legal claims" (Brand 1999, 46). Environmental INGOs may also directly undertake environmental protection projects. For instance in Zom, Senegal, INGOs and NGOs promoted environmental policies to protect the agricultural fertility and conserve topsoil (Wapner 2000, 91). Similarly, Jänicke and Weidner (1996, 305) found that in China and Russia, environmental INGOs and NGOs were "a basic precondition for better environmental performance."

Third, INGOs carry out a great deal of work in the international realm, and thus can be seen as a key part of the global regime. INGOs help define environmental problems and increase public awareness about the problems (Princen 1994). These efforts at publicity can create international public interest and change the position of states themselves (Wapner 2000, 95). INGOs are sometimes directly involved in the creation of treaty structures and laws, marshaling scientific research, or even administering the treaty (Princen 1994, 34). In many cases, INGOs carry out their own scientific research, or bring together scientific experts (Princen 1994, 34; Breitmeier and Rittberger 2000, 146). INGOs also contribute to the monitoring, verification, and enforcement of international treaty structures (Wapner 2000, 96). As Breitmeier and Rittberger (2000, 146) comment, INGOs and "NGOs provide unpaid services to, or carry out commissioned work for, international organizations or national governments."

World society scholars also use INGOs as an indirect measure of the overall Bee Swarm and its influence on particular national societies. Country-level memberships in INGOs are often used as an indicator of the penetration of international influences into the government and society of particular nations. One of the difficulties of testing the Bee Swarm model of change described in Chapter 1 is finding indicators of the multifaceted influences that come to bear on societies. It is challenging to gather data on the wide variety of mechanisms of social change, many of which are weakly related to any given outcome and highly correlated with other linkages. Country-level ties to INGOs are one easily measured proxy for the multiplicity of international influences on a given society. In some instances, world society scholars have collected specific measures of INGO activity in a particular domain, such as the environment or human rights. In other cases, researchers have used general measures of INGOs of all sorts – lumping environmental groups with groups devoted to topics ranging from recreation and sports to economic development to women's issues. While most of these eclectic organizations have no direct effect on environmental causes, the measure is nevertheless a potentially useful proxy. Those states that are deeply engaged in international global discourses on one issue – such as women's rights – also tend to be energetic participants in other issue-areas (Ramirez et al. 1997). As Wapner (2000, 92) argues, "Taken together, the host of environmental NGOs throughout the world represent a variegated presence through which voices and pressures in favour of environmental protection are being articulated and generated."

INGO memberships have become the standard indicator of world society processes, and proved extraordinarily successful in predicting policy diffusion and substantive change in a wide variety of domains. However, the influence of INGOs should not be taken literally. These results should not be interpreted as suggesting that INGOs are extremely powerful. INGOs rarely have the resources of leverage to bully or strong-arm national governments into submission to global norms. Instead, the presence of INGOs should be understood as the tip of the iceberg – the degree to which a nation-state is submerged in an ocean of world society cultural meanings, as discussed in Chapter 5.

Institutional Activity

Actor-centric theories assume that social structures are like a set of railroad tracks, on which agents travel on a highly constrained route. Activity within the constraints of social structure merely reproduces existing conditions and cannot bring about change. Actors are those who jump off the train between stations, to blaze their own trail in the wilderness. However, the emphasis on a heroic actor neglects the ability of agents to contribute to major social changes, operating within the context of existing institutional structures. Social changes may engage the efforts of thousands of bureaucrats and organizations and millions of ordinary people who are making mundane choices to implement laws, buy organic food, or simply turn out the lights. These effects can be dramatic in the aggregate.

From the world society perspective, institutional structures may be viewed as akin to a network of streets. Travel destinations are constrained – few cars opt to drive off the road for unmapped destinations. However, considerable ingenuity may be needed to maneuver around the grid of streets – as anyone who has gotten lost in Boston may agree. This ingenuity may be seen as a form of agency. Agents select destinations, pick routes, and cope with detours or obstacles that may arise. Much greater traffic can be accommodated within the road network than if each driver had to blaze her own trail. However, the incremental actions of various agents may slowly establish new streets to allow travel to new destinations, greatly increasing traffic to those places.

Agents often display creative interpretation, skillful translation, and resourceful innovation in their pursuit of institutional directives (Fligstein 2001). These efforts are facilitated by the support of institutional structures, which allow achievements that would not otherwise have been possible. The following section outlines three areas of activity in which world society scholars have identified agents working within institutional structures: (1) interpretation, (2) diffusion and translation, and (3) innovation.

Interpretation

Institutional structures may look solid from afar, but when examined closely they are skeletal frameworks that must be fleshed out. Formal laws are highly abstract, leaving the particulars of interpretation to government agencies, bureaucrats, and the judicial system (R. Andrews 2006; Dobbin 2009). Social norms may be even vaguer, creating imperatives for complex goals, including environmental protection and human rights, without specifying exactly what they entail (Hallett and Ventresca 2006). Ultimately, institutional structures require interpretation and implementation by agents.

Interpretation and implementation is no small task. Frank Dobbin (2009) has shown that policymakers typically pass broad legislation with little guidance on how the legislation should be implemented. Agents within governmental or organizational bureaucracies undertake the task of crafting specific interpretations and procedures to address regulatory requirements. In the case of U.S. affirmative action policies, these bureaucratic agents tended to be the human relations employees of corporations. Over time, the vague statutory wording of the 1972 Equal Employment Opportunity Act became rationalized into specific forms and requirements that have become the standards for affirmative action.

Hallett and Ventresca (2006) similarly show that broad social directives require significant interpretive work on the part of institutional agents. In this case, the supervisory management board of a gypsum mine provided a broad directive for increased rationalization and professionalization in the organizational procedures of the mine. Managers of the gypsum mine variously interpreted the directive of rationalization and utilized different mechanisms. Each boss had his own peccadilloes and varied in the effectiveness with which he complied with the directive for greater rationalization. Yet eventually, the organizational procedures of the mine become more rationalized in keeping with the broader structural mandates, albeit in a loosely coupled and haphazard fashion that reflected individual differences of interpretation.

Similarly, Strang and Meyer (1993) have argued that through the efforts of agents, procedures become standardized and routinized. Organizations create "standard operating procedures" that institutionalize environmental regulations. Such tendencies explain why environmental laws tend to result in general solutions rather than remedies tailored to the specific conditions of a particular local ecosystem (Hironaka 2002). Standardization – the emergence of policy packages – facilitates widespread diffusion of practices rather than each locale being forced to reinvent the environmental wheel (Strang and Meyer 1993).

The implementation of institutionalized rules often yields highly rationalized environmental policies and procedures that may not be truly optimal or effective in promoting environmental change. Indeed scholars have observed that corporate environmental policies are often formulated to avoid penalties or prevent litigation rather than to improve environmental outcomes (R. Andrews 2006). However, the world society perspective emphasizes the range of environmental agents and activities needed to bring even weak and ineffective policies into being. And as discussed in Chapter 3, the implementation of weak policy structures may still play an important role in social change, creating a workspace and bootstrapping additional pressures for environmental reform.

Diffusion and Translation

World society scholars have produced a great deal of empirical research showing that international discourses and policy packages spread to countries around the globe (Suchman and Eyre 1992; Strang and Meyer 1993). Critics have characterized diffusion studies as lacking attention to agency, implying that nations robotically adopt the mandates of a monolithic international structure. Yet the variability in national political contexts means agents must creatively translate abstract principles or norms in order for a policy to be adopted into a national political context (Westney 1987; Boyle 2002). Moreover, agents often play a central role in broadcasting or transmitting new cultural models worldwide.

First, agents must adapt global mandates to variations in economic and political conditions in each country. Adherence to the global mandate may be demonstrable in principle, but considerable heterogeneity might be evident in practices on the ground. Frank Dobbin (1994) has shown that national differences in economic and political structure require agents to adapt the same institution differently depending on the context. In the case of the establishment of national railroad systems, nearly every European state adopted a railroad system in the late nineteenth century. Yet each railroad system was adapted to the varied economic infrastructures and cultures of diverse nation-states. In France, the state created and maintained a centralized national railroad system, while in the United States, private corporations competed to buy up land and run their own railroads. Agents played a critical role in this adaption of the requirements of railroads to the specific national economic context.

Second, agents must translate Western-oriented packages for non-Western societies with different cultural traditions. Elizabeth Boyle (Boyle 2002; Boyle and Preves 2000) has shown that cultural differences affect the diffusion of international ideas to the local level in the case of female genital cutting. While a strong normative agreement against the practice of female genital cutting is evident in the international community, such policies have been domestically controversial in certain Islamic countries such as Egypt or Sierra Leone. In these countries, traditional cultural practices have resulted in sizable constituencies in favor of female genital cutting practices. Institutional agents must be particularly adept in such circumstances, in order to reconcile influential constituencies at both international and domestic levels with diametrically opposed viewpoints. Under these conditions, effective translation requires considerable savvy on the part of agents.

Third, institutional agents often play a central role in transmitting new cultural meanings and discourses and educating people at the national and subnational levels. Depending on the environmental issue at hand, this may be a major endeavor. For instance, in the case of hazardous waste in the United States, thousands of waste management facilities and millions of hazardous waste producers needed to be recruited and reoriented toward new approaches for dealing with hazardous waste (Mazmanian and Morell 1992). Separate from the issue of overcoming corporate or individual interests to pollute, an issue taken up in Chapter 5, translating information about new environmental regulations or improved practices is a significant task that requires substantial numbers of agents.

Agents vary in their skill in translation, as well as in the resources and political will available. Consequently, the diffusion of institutional structures to the national or subnational levels is likely to display considerable heterogeneity on the ground. This heterogeneity does not necessarily reflect opposition to the global environmental regime, as discussed in Chapter 1. Instead, heterogeneity reflects the diversity of local conditions and distinctive outcomes of the process of translation.

Innovation

From the actor-centric perspective, the concept of innovation holds out the hope for substantive change from within an institutional structure. For instance, Dimaggio (1988) develops the concept of the "institutional entrepreneur," an individual actor motivated by intrinsic self-interest who is able to deviate from existing structural constraints. Where social movement scholars look to activists to heroically initiate social change, organizational scholars have looked to the entrepreneur.

However, innovations that run against the grain of preexisting institutional structures often fail, as discussed in Chapter 2. Innovation is more likely to be successful when agents work within structures by seeking to incrementally expand existing institutions and meanings – expanding them to a new venue, modifying them slightly, or forging new links between two existing structures (Fligstein 1997; Lounsbury and Crumley 2007). These innovations do not occur automatically, but require considerable effort by agents. Nor are the outcomes of these efforts preordained. Agents vary in their capabilities and visions, which affect the form and success of innovation.

One common type of innovation is the expansion of existing environmental structures into new domains. Environmental agents, in their enthusiasm for environmental protection, seek to curtail an ever-growing range of pollutants in an ever-expanding number of domains. At the international level, expansion might imply including a new form of air pollutant into the Long-Range Transboundary Air Pollution Convention. At the micro level, expansion might imply setting up a recycling program in an additional community or workplace. Although these expansions may appear straightforward, they should not be considered automatic or inevitable. Instead, such innovations often require considerable effort and ingenuity on the part of agents.

A second form of innovation is the linking of one institutional structure to another. Organizational theorists have focused primarily on network linkages of resources, information, or personnel between institutional structures (Burt 1992; Granovetter 1973; Mizruchi and Galaskiewicz 1993). However, these linkages often require innovation in cultural discourse and meaning to reconcile previously independent institutions as well. For instance in the 1970s and early 1980s, the World Bank's primary goal of third world economic development was largely disconnected from environmental logics and discourses. With the expansion of environmental cultural meanings by the mid-1980s, however, the World Bank re-constructed its logic as congruent with new environmental logics of environmentally sustainable economic development (Goldman 2005; Fox and Brown 1988; Le Prestre 1989; Nakayama 2000). These processes of the reconstruction of social meaning are taken up in Chapter 5.

A third form of innovation is the construction of new environmental problems that require social actions and institutional structures to address. The identification of a new type of environmental degradation generally requires significant scientific effort to discover in the first place, and even more effort to substantiate on the scale needed for national and international policymaking (see Chapter 3). Moreover, each environmental issue must contend for recognition on the global political agenda, which poses additional obstacles (discussed in Chapter 2). Placing a new environmental problem on the global agenda is a major achievement that involves the efforts of hundreds or thousands of different environmental agents.

These innovations result in new cultural meanings and new patterns of activity. Agents, often zealous on behalf of environmental institutions, continually seek out new problems, new venues, and create new structures to expand the protection of the environment. Nevertheless, those looking for more revolutionary kinds of social change may see such activity as mundane. In the long run, however, these humdrum forms of innovation lead to dramatic social changes. Indeed, such changes generate the seemingly unstoppable structures against which charismatic actors of future generations may oppose.

The Problem of Hazardous Waste

The case of hazardous waste disposal in the United States highlights the difference in effectiveness between actors and agents. The concepts of "actor" and "agent" represent ideal types. For the purposes of this example, the grassroots social movements protesting hazardous waste facilities in the 1980s will be characterized as "actors." In contrast, the army of bureaucrats, waste management facility operators, and even some waste producers are the "agents." Although the "actors" of the NIMBY movement were spectacularly successful in preventing the development of new waste management facilities in the 1980s, they largely failed to address the broader problem of environmentally safe waste disposal. By contrast, despite early failures, the "agents" incrementally created a waste management regime in the United States that has significantly improved practices over the past few decades.

As in many environmental arenas, social movement actors operating outside the scope of conventional political and organizational life have had only a limited effect in improving environmental outcomes. In the case of hazardous waste disposal, neighborhood social movement groups were mainly successful in preventing nearby waste treatment plants from being built. However, such efforts only shuffled the problem of waste disposal to less mobilized communities. Nor did these actions do much to address the exponential increase in the production of hazardous waste that was occurring simultaneously.

By contrast, a variety of bureaucratic and organizational agents slowly constructed a national infrastructure for the disposal of hazardous waste over the same period. Agents working within government institutions have innovated by expanding bureaucratic structures for hazardous waste disposal, implemented legislative requirements, and diffused these mandates to states and municipalities. Despite numerous failures, such as the highly publicized fiascos of Superfund, the United States has greatly improved its handling of toxic waste in the past thirty years.

Actors Mobilize against Hazardous Waste

Citizen activism against hazardous waste facilities in the United States is considered one of the most successful environmental social movements in the modern period (Szasz 1994; Mazmanian and Morell 1992; Pellow 2007). Beginning in the late 1970s, neighborhood groups across the United States mobilized under the slogan "Not In My Back Yard" to protest proposals for landfills, waste incinerators, and hazardous waste storage. The movement began with middle-class citizens mobilizing to protect property values, but later engaged a broader range of participants including lower-income neighborhoods without a historical propensity for social movement activity (Szasz 1994).

Despite its successes, the hazardous waste movement illustrates the challenges that actors face as they seek to bring about large-scale environmental change. Until the widely publicized events of Love Canal in 1979, the U.S. public was largely unconcerned about the effects of toxic waste. The media explosion around Love Canal galvanized activity by neighborhood grassroots groups as well as by local and federal authorities. However, grassroots activists lacked the organizational capacity to broaden to other issues or join with other groups. Instead, these local groups faded as the urgency of the issue diminished. While groups were often successful in blocking new waste treatment plants, such action did not aggregate to a broader environmental solution for waste disposal problems.

Lack of Interest: The issue of hazardous waste exploded into national awareness with the shocking media coverage of Love Canal in 1978.[1] The community in Love Canal, New York, had been built on top of an industrial waste site that had been closed in the 1950s. For two decades, the residents of Love Canal had been suspicious of the mysterious black sludge that seeped into cellars and covered playgrounds (Walsh, Warland, and Smith 1997). Citizens were concerned about the high rates of birth defects and cancer that seemed greater in their neighborhood than in other areas of the state. An enterprising journalist exposed the issue to a wide audience for the first time. Andrew Szasz (1994, 34) describes the result: "Practically overnight, hazardous waste went from being a hazy, poorly organized perceptual object in popular imagination to being one of the most feared of environmental threats."

Prior to Love Canal, social movement activists had been unable to stimulate concern among the American public. Scholarship has shown that grassroots organizations had been mobilizing to protest hazardous waste disposal throughout the 1970s and even earlier (Szasz 1994). Yet such activity had little effect on public awareness. Hazardous waste had not yet been identified as an issue on the public agenda, nor had options for its disposal been defined. By 1973, an Environmental Protection Agency survey found that "people still viewed the prospect of having a toxic waste facility for a neighbor with equanimity" (Szasz 1994, 41). Szasz (1994, 70) concludes, "Contamination protests before 1978 [Love Canal] were slow to develop because activists found it hard to convince the larger community that there was, indeed, a real problem."

---

[1] In this account, the media explosion surrounding Love Canal in 1979 may initially appear to be an "exogenous shock" that affected both social movements and institutional structures in the United States. Yet the public reception of Love Canal depended upon preexisting institutional activity related to public health over prior decades, not to mention the growing national-level and global environmental regimes, which legitimated and provided cultural building blocks for the construction of environmental and health problems in the public sphere.

Lack of Organizational Resources: One of the striking aspects of the hazardous waste social movement was its true grassroots nature. Scholars have found that middle class activists are typically the most effective organizers for social movements, even if the movement is on behalf of oppressed or impoverished groups (Della Porta and Diani 2006). In contrast, the hazardous waste movement in the United States was successful in mobilizing neighborhoods at all socioeconomic levels, including poor neighborhoods with little history of activism (Szasz 1994; Pellow 2007).

Yet the U.S. hazardous waste movement generally failed to build broader institutional structures beyond the neighborhood group. A few umbrella social movement organizations arose, such as the Citizens Clearinghouse for Hazardous Wastes, which became a national coordinating body for the movement in the early 1980s (Szasz 1994; Walsh et al. 1997). Some evidence of limited networking and issue development has also been noted (Szasz 1994, 71–72). Yet for the most part, the hazardous waste grassroots organizations did not create networks that united various neighborhoods under a common banner. Instead, most of these movements died out once their local aims were achieved.

Thus the hazardous waste movement failed to follow in the footsteps of other successful social movements. Previous movements such as women's rights or the African-American movement were able to bootstrap local movements into a national activist structure, headed by organizations such as the National Organization for Women (NOW) and the National Association for the Advancement of Colored People (NAACP). In contrast, hazardous waste activists rarely addressed environmental problems beyond the scope of a single particular neighborhood in most instances.

Impact on Environmental Outcomes: The explosion of public interest generated a large number of social movements throughout the 1980s. Grassroots organizations protesting "Not In My Back Yard" (NIMBY) forced waste plant developers to abandon their plans in a large number of U.S. cities. Waste disposal problems could not easily be dumped on less economically advantaged communities, since poor neighborhoods mobilized as well. Indeed, it became difficult to develop any new sites for waste disposal, as nearly every new waste disposal plant drew protest by the mid-1980s. Walsh et al. (1997, 1) reported that by 1990 the United States only had 140 trash incinerators operating because nearly twice that number had been canceled in the previous decade due to "grassroots opposition [that] made it increasingly difficult to build a trash plant anywhere in the nation after 1985."

Such resistance is understandable and even admirable from a local point of view. However, citizen opposition did not stem the exponentially increasing levels of garbage being produced by American consumers. Ironically, these movements often blocked the development of safer and more environmentally friendly garbage treatment facilities. Although older landfills usually lacked the environmental safeguards that were required for newer landfills, local resident protests blocked the building of new and environmentally improved landfills. Consequently the continued operation of older landfills "largely stymied the push for a more rapid transition to new treatment technologies" (Mazmanian and Morell 1992, 106). Citizen protests delayed implementation of hazardous waste policy that in many cases arguably led to increased health and environmental risks or at least increased the cost of hazardous waste disposal (Kraft and Kraut 1988).

In sum, grassroots protests did little to solve the broader problem of hazardous waste disposal, despite success in keeping waste disposal sites out of particular neighborhoods. Protests generally did not target the production of hazardous waste, which was occurring in the United States at increasingly prodigious levels. Nor did these movements push for healthier and more ecologically friendly solutions. As Mazmanian and Morell (1992, 181) conclude, "The simple answer to why new hazardous wastes facilities ... have not been developed throughout the first toxics decade in the United States is local opposition."

Agents and the Expansion of Waste
Management Infrastructure

The public spectacle of Love Canal also galvanized a wide range of agents within the bounds of preexisting institutional structures. Hazardous waste became a prominent issue on the political agenda in response to the increased public outrage. Congress authorized the Environmental Protection Agency to set up Superfund in 1980 to clean up particularly toxic sites. Greater regulation was developed for the disposal of toxic wastes, incentives were created to improve disposal technologies, and additional data were collected on hazardous waste production and disposal. All of these activities required effort on the part of government officials, operators of landfill and incineration facilities, and even the producers of hazardous waste. Consequently, despite the exponential growth in the production of hazardous waste, disposal in the United States has greatly improved over the past thirty years. This section discusses the innovation, implementation, and translation efforts carried out by agents to improve the handling of hazardous wastes in the United States and worldwide.

Innovation: New institutional structures are usually developed as extensions of existing structures, as discussed in Chapter 2. This was definitely the case for hazardous waste disposal in the United States. While publicized media events such as Love Canal and the 1984 Bhopal disaster in India provided initial spurs, these innovations built upon existing infrastructures. Actor-centric accounts often overlook the work that goes into these mundane forms of structural innovation. Extending an existing road may appear commonplace, yet allows the expansion of traffic to new destinations. Similarly, these mundane forms of innovation provided the structural basis for significant shifts in U.S. hazardous waste disposal practices.

In their plodding way, legislators in the American government had already created federal policy regulating hazardous waste prior to the events at Love Canal. Passed in 1976, the Resource Conservation and Recovery Act (RCRA) regulated types of waste that could be disposed in a landfill. Almost as an afterthought, the unregulated dumping of identified hazardous substances was prohibited in Subtitle C. This expansion of federal authority to include solid waste disposal was viewed as simply another bureaucratic regulation in the ever-expanding government arsenal, arousing little public or corporate attention. As one scholar notes, "minimal opposition to RCRA" was raised "on the part of generator industries" (H. Barnett 1994, 56). Nor were citizen activists responsible for initiating the law. As social movement scholar Andrew Szasz (1994, 15) puzzles, "If there was no public opinion and no social movement, if there was *no issue* yet, how did Congress come to regulate hazardous waste?" He answers his own question by concluding that the RCRA was driven by "larger developments in national politics" (Szasz 1994, 15).

The public shock resulting from the Love Canal scandal forced government legislators and EPA officials to develop further innovations to U.S. waste disposal policy. These innovations extended existing environmental structures within RCRA and the Environmental Protection Agency (EPA) and also created the new program of Superfund within EPA jurisdiction. While RCRA had been written to prevent the unregulated dumping of hazardous waste in the present, it had not yet been recognized that previous unregulated dumping had created a toxic waste legacy that would require federally mandated amelioration (H. Barnett 1994, 56). In response, Congress passed new legislation in 1980, known as CERCLA, which generated even more bureaucratic requirements, raised standards for waste incineration, and authorized the Superfund program.

In 1984, the chemical plant accident in Bhopal, India provided another opportunity for institutional innovation in U.S. hazardous waste policy. In 1986 the Emergency Planning and Community Right-to-Know Act of 1986 (EPCRA) mandated the annual reporting of toxic chemical releases and transfers for a defined list of chemicals, industries, and facilities (Hamilton 2005, 5). These data were tracked and reported by a database labeled the Toxics Release Inventory. Interestingly, the public reporting of these data produced effects consonant with the effects of structure discussed in Chapter 3. When the first figures from the Toxics Release Inventory were released in 1989, companies pledged to reduce toxic emissions and environmentalists publicized the numbers to bring about change. By 2001, total releases and transfers had dropped 54.5 percent for those chemicals and industries in the database, despite the lack of government mandates (Hamilton 2005, 5).

These innovations in bureaucratic structures may not appear to be sufficiently radical from an actor-centric point of view. However, these new laws and programs involved the efforts of thousands of agents to prepare the legislation, implement the regulations and programs, and communicate with the public. Such efforts may go unnoticed in the search for revolutionary change. Nevertheless, these forms of mundane innovative activity can bring about lasting change in environmental outcomes, whereas spectacular bursts of action subside without much consequence.

Interpretation: Consistent with the Bee Swarm model, regulatory structures addressing hazardous waste have been only weakly linked to substantive improvements in waste disposal. Superfund in particular was widely condemned, although many other aspects of U.S. hazardous waste policy have been criticized as well. But as argued in Chapter 3, even inadequate institutional structures can bootstrap eventual improvement in environmental outcomes. This section underscores the multitudes of agents involved in implementing these programs that provided the foundation for U.S. waste management institutions.

Although hazardous waste legislation predated the Love Canal scandal, six years passed before the EPA managed to issue regulations for handling toxic waste for landfills and storage ponds as mandated by RCRA. Regulations for incinerators and other types of facilities would not be developed for several more years (Mazmanian and Morell 1992, 91). As Mazmanian and Morell (1992, 78) mourn, "RCRA sank almost without a trace for half a decade into the mire of the already overworked

EPA bureaucracy." Scholars have assumed that the public pressure generated by Love Canal in 1978 was required for implementation (Szasz 1994; Mazmanian and Morell 1992).

It took prodding by several environmental-group lawsuits, four years of study and testing, negotiation with industry, and internal wrangling before EPA came up with its operational definition of hazardous waste, along with the testing protocol industry was to follow and the first lists of 450 substances and several dozen industrial waste streams that would ultimately fall under the RCRA regulatory umbrella. (Mazmanian and Morell 1992, 81)

Yet even this disparaging commentary hints at the magnitude of the task mandated for EPA in 1976. The implementation of the first federal policy regulating hazardous waste started from scratch. EPA employees were required to generate a list of toxic chemicals to be regulated, out of the hundreds of thousands of chemicals produced daily in the United States, few of which had ever been tested for toxicity. These chemicals were emitted by at least 20,000 major sources of air pollution and 68,000 sources of water pollution (Vogel and Kessler 1998, 24). Data had to be established for the amounts of hazardous waste produced and disposed, which was eventually estimated at somewhere within the wide range of 145 to 435 million tons annually (Mazmanian and Morell 1992, 83). Moreover, EPA was responsible for the implementation of new regulations for the 5,000 facilities for waste disposal (Mazmanian and Morell 1992, 83). Such activity required the coordinated efforts not only of EPA employees, but state officials, industry experts, and waste disposal facilities. Given the magnitude of the task, it is perhaps unsurprising that it took six years for regulations to be developed.

Superfund was to be the target of even more enraged criticism from a public mobilized by Love Canal and NIMBY protests. Once again, EPA was placed on the hot seat to interpret CERCLA and develop regulations that would clean up an unknown number of Love Canal–type sites throughout the country. As usual, the legislators of CERCLA had left the regulators at the EPA to define the standard for cleanup. "How clean is clean?" muse Mazmanian and Morell (1992, 42). EPA was given discretion to interpret the ambiguous legislative statutes, with contradictory stances between local residents wanting zero contamination on the one hand to industries that preferred minimal costs on the other (Mazmanian and Morell 1992, 43). By 1990 (ten years after CERCLA), the EPA had managed to complete cleanup on only a few sites, encouraging scholars to declare, "Superfund had become a superfailure" (Mazmanian and Morell 1992, 30).

Nevertheless, by 1984, EPA had developed a list of 786 proposed sites requiring cleanup on the National Priorities List. In addition, an estimated 2,000 sites might eventually require cleanup, and between 131,000 to 379,000 sites were considered of "potential concern" (H. Barnett 1994, 27). It may have been that the response of the EPA was inadequate, as some scholars have argued, even given the magnitude of the task. However, even partial implementation involved hordes of environmental agents. As Szasz (1994, 139) remarks, "The very presence of a regulatory statute fosters the development of 'issue infrastructure,' that complex of knowledge, technology, and institutions that makes it possible for any society to understand and cope with any issue." The EPA was tasked with the creation of a hazardous waste regime essentially from the ground up in 1976. These efforts fell short – indeed, the magnitude of the task nearly insured inadequate implementation. Yet these efforts at implementation did succeed in creating a multitude of agents concerned with hazardous waste disposal where few had existed previously.

Ultimately, a great deal of social change has transpired in the U.S. management of hazardous waste. As Mazmanian and Morell (1992, 7) conclude, "Many significant changes in toxics management have in fact occurred. In several cases, industries have shown a remarkable willingness to reduce their waste streams, which they had simply been dumping into landfills, toxic pits, or lagoons." Even social movement scholar Szasz agrees that there has been substantive improvement:

The past fifteen years has also seen the development of a real disposal and treatment industry. ... Today, the United States has a modern, multibillion-dollar disposal industry. Although it is true that even the biggest and best firms in the industry have poor operating records ... the overall trend is positive. The worst of the old facilities are closing, going out of business. An active research program is developing new and promising treatment technologies. (1994, 139)

Translation to Other Countries: An international framework for hazardous waste disposal is also slowly emerging. The first major milestone was the 1989 Basel Convention on the Transboundary Movement of Hazardous Wastes and Their Disposal and the subsequent amendment of 1995 that banned exports of hazardous waste from industrialized states to non-OECD countries (Clapp 2001). Yet this is but a small step in developing an infrastructure for handling hazardous waste – essentially NIMBY on a global scale. The Basel Ban Amendment seeks to keep hazardous waste within the OECD countries, yet increasing numbers of non-OECD countries are beginning to produce their own hazardous wastes domestically. These newly industrializing countries often do not have the local capability to handle hazardous waste in an environmentally safe manner. However, the development of such institutional structures tends to occur in a slow and mundane fashion, far from the excitement of media events.

The worldwide establishment of hazardous waste programs appears to be in the early stages of a typical diffusion pattern, of the sort frequently observed by world society researchers, although the process is not far along. Countries highly embedded in world society, such as the United States, Denmark, Canada, and Germany, created hazardous waste legislation in the 1970s[2]. Countries that are less embedded in world society, such as Malaysia, Hong Kong, Thailand, and Indonesia, followed in their footsteps during the 1990s.[3] However, this diffusion of hazardous waste legislation to the newly industrialized countries should not be interpreted as a passive process. Each government must deal with variations in geography, differences in the types of hazardous waste produced, the prior infrastructure for dealing with waste, and the culture context of a given society. As Probst and Beierle (1999, 5) conclude, "Developing a [hazardous waste management] system is a complicated undertaking, and the path taken often depends on the particular geography, demographics, industrial profile, politics, and culture of a country."

Many European countries began to develop a hazardous waste infrastructure within the 1970s and early 1980s. However even in Europe, scholars noted that "the road to establishing an effective hazardous waste management (HWM) program is a bumpy one" (Probst and Beierle 1999, 1). Environmental agents were needed to adapt hazardous waste structures to historical and geographic differences in the waste management structures in each country. For instance, the United Kingdom had historically focused on the development of landfills as its principal means of waste disposal, while other countries such as Denmark and Sweden had primarily emphasized incineration (Mangun 1988, 210). Countries created their own lists and standards for toxic chemicals, "which resulted in substantially different policies in each of the member countries" (Mangun 1988, 209). Although the European Economic Community passed a directive in 1978 that required member states to "take the necessary steps to dispose of toxic and dangerous waste without endangering human health

[2] Germany 1972, Denmark 1973, United States 1976, and Canada in 1980 (Probst and Beierle 1999, 10).
[3] Malaysia 1989, Hong Kong 1991, Thailand 1992, and Indonesia in 1994 (Probst and Beierle 1999, 10).

and without harming the environment," only six states had complied by 1983 (Mangun 1988, 209). Probst and Beierle (1999) find that countries generally require ten to fifteen years to implement a reasonably effective hazardous waste regime. As late as 1988, one scholar commented, "the disposal of hazardous waste continues to remain a major difficulty for all Western European countries" (Mangun 1988, 214).

Diffusion of hazardous waste regimes to less developed countries is additionally hampered by the considerable work needed to adapt regulations to the specific situations of each country. In contrast to the relative predictability of hazardous waste production in industrialized economies, the small-scale industries in developing countries tend to produce heterogeneous amounts and types of hazardous waste that have been difficult to monitor and handle appropriately (Kummer 1994, 11). In Hong Kong for instance, over 90 percent of hazardous waste producers were small businesses located in high-rise buildings that housed other inhabitants, rendering on-site treatment of hazardous waste unrealistic (Probst and Beierle 1999, 78). Malaysia faced difficulties such as "a lack of trained personnel, a lack of facilities, and the difficulty of preparing schedules for toxic and hazardous wastes," and opened a comprehensive waste management facility nearly a decade after its initial hazardous waste legislation (Probst and Beierle 1999, 73, 75). In Thailand, experts have opined: "there is some question of whether the expertise exists within any of these [government environmental] agencies to implement a comprehensive hazardous waste management system" (Probst and Beierle 1999, 87).

Diffusion of structures such as hazardous waste legislation might appear a passive process from the outside. Yet adaptation of a regime to a different economic, political, and cultural system requires a great deal of effort by environmental agents. As discussed in Chapter 3, it is not surprising that legislative requirements have not had a simple linear effect on improved outcomes. Yet, the short-term ineffectiveness of these structures does not mean that the overall Bee Swarm of pressures will be inconsequential. The development of institutional structures, such as international treaties or national legislation, enables and empowers agents to be more effective. Over time, the continued efforts of pro-environmental agents tend to cumulate into improved environmental outcomes.

Conclusion

The world society perspective identifies institutional structures as the source of social change. Institutions serve to construct new social problems and establish workspaces in which they can be addressed. However, the existence of a structural workspace is insufficient without agents to carry out the work – collecting and assembling data, devising regulations, transmitting policy packages, and so on. Theorists looking for more revolutionary actors to bring about major social upheaval may overlook the activities of these less glamorous environmental agents. However, social change is rarely the result of revolution – indeed, scholars have lamented how frequently revolutions fail or post-revolutionary societies replicate the conservative structures they sought to replace. Rather than focusing on heroic actors that radically oppose existing structures, world society theory focuses on agents – people and organizations that are empowered by, and work within, the cultural frameworks and regulatory structures of institutions.

One might contrast the activities of vigilantes compared with a police force. Vigilantes work outside of legal institutions and in particular cases may be instrumental in bringing justice to cases that the law either cannot or will not touch. Yet the bulk of law enforcement falls to the police, who act as agents of criminal justice institutional structures. There may be many flaws with the criminal justice institution – the prevalence of its agents should not be taken to imply that the institution is a fair or optimal one. However, if the question is effectiveness, it seems clear that the police are far more effective than vigilantes in carrying out the mandates of legal institutions. While vigilantes may be more effective in particular cases, they cannot match the police in the sheer bulk of criminals processed by legal institutional structures.

Agents are central to institutional explanations for change. And, the concept of the agent helps clarify the meaning of agency within world society theory. Ideal typical "actors" who blaze new trails face tremendous obstacles, and rarely succeed in transforming society as a whole. Agents, operating under the aegis of institutional regulatory structures and cultural discourses, often prove surprisingly effective in the long term. Institutions are sufficiently skeletal and loosely defined that agents can exercise a great deal of initiative and discretion. Through processes of interpretation, translation, and innovation, agents can be very consequential, bringing contingency – and a sense of agency – into world society explanations.

5

Cultural Meaning

World society theory is at heart a cultural theory. As John Meyer notes, "environmental concerns are matters both of social organization and of an embedded culture and set of meanings. We talk differently about the air, water, earth, and biosystem than we used to, and perceive many detailed problems and crises" (J. Meyer 2002, xiii). The changed meanings of environmental issues pervade national and international discourses and provide an essential motivation for social change. But cultural meanings, and their role in social change, can be hard to describe and pin down and so may be bypassed by scholars or subjected to airy waves of the hand. This chapter examines the concept of culture within world society theory. Ultimately, culture is characterized as a thick set of meanings embedded within institutions, and an integral part of the overall Bee Swarm process of social change.

Scholarship on international regimes often invokes a relatively thin conception of culture as specific norms or rules that can evoke either compliance or evasion depending on convenience. Norms promoting recycling or polite use of cell phones may be socially acknowledged yet are easily breached when expedient. Similarly, corporations and governments may pay homage to environmental norms in their rhetoric yet ignore these norms when advantageous to bottom line profits or strategic interests. From this perspective, material and political interests are likely to undermine the effect of pro-environmental norms when interests and norms are in conflict.

This "thin norm" understanding neglects the emergence of a whole new cultural understanding of environmental issues. The meaning and salience of the natural environment has undergone a sea change with the expansion of the global environmental regime. Obscure and distant environmental concerns – the Amazonian rainforest, the ozone hole in the Arctic, or endangered species in Africa – have become part of contemporary social reality, even for people who are directly unaffected by those particular environmental problems. Changes in meaning and the increasing cultural salience of the environment provide a foundation for increased activity that addresses environmental problems.

This approach provides a new lens for understanding opposition and conflict surrounding environmental issues. Cultural understandings in environmental meaning do more than simply constrain or shame actors. Over time, shifts in cultural meaning, along with other institutional changes, can fundamentally reshape the interests of states and firms. Issues that appeared intractable at one point in time, because of powerful corporate opposition, melt away in subsequent decades, and new arenas of contestation and conflict emerge. Conflict goes hand in hand with the expansion of the global environmental regime, and entrenched opposition at the outset may not prove to be a barrier to long-term pro-environmental change.

This chapter unpacks the cultural dimensions of international environmental institutions and explores the implications for social change. The chapter begins by contrasting a view of culture as "thin" regulative norms compared with a more useful view of culture as "thick" meanings. The chapter theorizes how new environmental meanings generate conflict and resistance to change, even as they reshape interests and encourage diffusion of pro-environmental policies and practices.

Finally, the chapter turns to the empirical case of the global climate change regime. Climate change is a challenging case, since the process of institution building is only beginning and the ultimate success of the regime is uncertain. Indeed, international climate change efforts have been widely criticized as a failure in which corporate interests largely won out (Newell 2000). This chapter argues that the nascent climate change regime is shifting cultural meanings and reshaping the interests of key players. Despite significant resistance to pro-environmental reforms, the global climate change regime is laying the groundwork for substantial social change in the future.

Sociological Perspectives on Culture

Culture is an all-encompassing medium that is difficult to pin down analytically – as pervasive as the sea to the fish that swim in it. One way to gain purchase on culture is to draw contrasts over lengthy periods of time. For instance, Zelizer's (1994) classic study shows that cultural perceptions of the value of children have transformed dramatically over the past century. Likewise, the meaning of cigarette smoking has changed a great deal in the past few decades. Not long ago smoking was a sign of sophistication, whereas today it is perceived in a more pejorative light.

The concept of culture is employed in very different ways across fields, and even within sociology (Lamont and Small 2008). The "thin" norms approach views culture as a set of normative rules that constrain behavior (Hechter and Opp 2001; Goertz 2003; Dimaggio 1988). Sociologists have long observed the power of norms in everyday interactions. Introductory sociology classes popularly require students to breach a norm in order to experience the social reactions as well as their own discomfort (Goffman 1974). While breaching norms may be uncomfortable in everyday settings, it is easy to see why corporations might evade norms when bottom-line profits are at risk. Anecdotes abound in which corporations pay lip service to environmental norms while continuing to violate norms in practice (Leonard 1988; Chatterjee and Finger 1994). International environmental institutions typically have very weak enforcement and sanctions, so the cost of violating international norms tends to be mild disapprobation at best. When environmental norms conflict with fundamental interests such as profit maximization, it is easy to see why interests win.

In contrast, culture might more usefully be conceptualized as a "thick" layer of social understandings consisting of ontology, salient meanings, and frames for action. From this perspective, culture is a central component of institutions, intertwined with the organizational structures and agents discussed in previous chapters. Institutional structures embody and stabilize cultural understandings of the environment that, at least in recent decades, support and empower a great deal of activity. World society scholars have traced changing cultural meanings over time, as reflected in the content or discourse of international treaties and organizations (see Frank 1997 and Frank 1999 for examples).

First, the "thick" conception highlights the importance of culture as ontology – basic social understandings about the world and what exists. Concepts such as acid rain, oil spills, or the ecosystem have become meaningful terms in society (Fourcade 2011). Contemporary understandings of the natural environment encompass an ever-greater range of human-nature interaction on land, air, and water, and a vast expansion of both scientific and lay knowledge about the environment. While contemporary cultural understandings of the environment sometimes draw on scientific ones, and are certainly legitimated by institutionalized science, cultural views cannot be reduced to simply learning more about objective reality. Instead, as discussed in Chapter 2, new environmental meanings evolved in a historically contingent manner with the growth of the global environmental regime. In many cases, cultural change predates or is only loosely connected to scientific consensus. In any case, an increasing number of environmental issues have been constructed and rendered meaningful and actionable over time. Climate change, for instance, is now an object of social discussion – a process that is linked to the emergence of global institutional structures.

A second aspect of cultural meaning might be termed its salience – its meaningfulness or apparent relevance to society or individuals. Climate change has been a topic of scientific discussion since the 1980s, but was not initially seen as broadly relevant to society outside of academic circles. In recent years, climate change has become *meaningful* to a global audience, increasingly embedded in economic, political, and social discourses. This does not imply unanimous agreement – indeed, there is ongoing conflict and contestation over the appropriate response and even the ontological existence of climate change. This widespread conflict, however, reflects the growing salience of climate change in contemporary world society.

Third, cultural meanings include frames or imperatives that motivate action. Social movement scholars have long argued that the adoption of cultural frames provides the basis for social action (Snow and Benford 1988; Ferree et al. 2002). If air pollution is viewed as an unfortunate but inevitable by-product of industrialization, collective action against pollution is not likely. Within this frame of resignation, Londoners in the Victorian era resorted to carrying black umbrellas in response to the sooty air that irremediably dirtied light colored clothing (Brimblecombe 1987). If, however, the concept of pollution connotes widespread negative health consequences and dying animals, antipollution mobilization becomes more feasible. Cultural meanings embedded in institutions promote and motivate patterns of social action.

Some environmental meanings become taken for granted. Concepts such as "pollution" and "hazardous waste" are now universally understood as pejorative. Other meanings, however, are contested. Within the United States, for instance, there is still debate whether "climate change" represents an urgent environmental problem or a scientifically flawed theory that will waste millions of dollars. Similarly, environmentalists, hikers, and paper companies contend over the cultural meaning of forests. This chapter explores the effect of "thick" cultural meanings of the environment on interests and competing logics that appear to limit pro-environmental action.

The Institutional Reconstruction of Interests

When international institutions are viewed as embodying "thin" norms or rules, it is hard to imagine how substantial social change will occur in the face of entrenched interests. While international norms may bring social disapproval or mild embarrassment to countries or firms that fail to comply, they generally lack potent sanctions. Under these conditions, immediate strategic or material interests seem likely to trump pro-environmental norms. Consequently, scholars (and environmentalist activists) have been skeptical that international norms will prompt states and firms to effectively address environmental problems.

However, the scholarly discussion of international norms frequently overlooks how economic interests evolve in response to institutions. Put simply, as institutional structures change, interests change. This shift is evident in conventional institutional analyses of corporate interests (Bess 2003; A. Hoffman 2001; Kamieniecki 2006). This chapter describes prior work on institutions and interests, and then adds the world society focus on cultural meaning. By highlighting the role of "thick" culture, the chapter offers a more complete explanation of how institutions transform interests and generate social change.

The Conventional View of Interests

Traditional environmental scholarship focuses on material interests as the primary driver of behavior (Hechter and Opp 2001; Goertz 2003). Environmental accounts begin with the parable of the commons, in which it is in the interest of each farmer to add one more sheep to the grazing commons, despite the knowledge that overloading the commons will negatively impact all of the farmers (Harding 1968). As Brenton (1994, 4) writes in less ovine terms, "Left to themselves, profit-maximizing individual users ... will expand economic activity, knowing that the consequent polluting emissions will be dispersed among all, while the economic benefit from the extra production will be confined to the producer alone." In short, economic interests will inexorably lead to environmental deterioration.

From this conventional viewpoint, interests exist *a priori* and strategies for furthering these interests are generally self-evident (Hechter and Opp 2001; Goertz 2003). Even neo-institutional scholars have, at times, employed the idea that interests are *a priori* motivators of behavior and are therefore unexplained by institutional processes (e.g., DiMaggio 1998). When actors are fundamentally motivated by self-interest, behavior is unlikely to change unless coerced by strong sanctions. Unfortunately, environmental policies and treaties rarely incorporate strong sanctions and enforcement (Brown Weiss 1998). States and corporations may pay lip service to norms of environmental protection but are likely to comply only when norms are congruent with calculations of self-interest.

Images of self-interested action are routinely invoked to explain resistance to environmental policies, regulations, and norms. The implementation of new environmental policies is often costly for firms, leading to short-term and possible long-term decreases in profit. According to one source, annual expenditures for pollution control tripled in the United States from 1972 to 1990 and increased an additional 50 percent by 2000, amounting to 2 percent of the U.S. gross domestic product (A. Hoffman 2000, 1). Production costs increased at least 5 percent for nearly all businesses in the United States because of environmental policies (Wubben 2000, 25). These costs are substantial, and there are many anecdotal accounts in which corporations have found it in their interest to seek exemptions or evade required implementation (Leonard 1988; Chatterjee and Finger 1994). For instance, Bess (2003) finds that industrial corporations in France in the 1980s were strongly opposed to environmental reforms and blocked many proposed legislative changes, in line with predictions of profit-making corporate interests.

However, there is reason to question this conventional perspective in which economic interests are implacably opposed to environmental protection. When scholars have examined corporate interests over time, interests often turn out to be less entrenched than initially assumed. Bess (2003) was surprised to find that twenty years later the same anti-environmental French corporations were not only complying with environmental regulations, but had become exemplary "green" companies. Andrew Hoffman (2001) found a similar pattern in U.S. corporate behavior. By the 1990s, U.S. corporations that had opposed environmental regulations twenty years previously were proactively implementing environmental behaviors into their operating procedures. When examined at a single point in time, environmental norms appear weak compared to the strength of entrenched interests. However, this static picture is misleading because interests are themselves transformed as institutions change.

Institutions and the Reconfiguration of Interests

Studies of environmental contestation typically focus on a single snapshot in time that allows laws, competitive pressures, and consumer preferences to be held analytically constant. However, the resulting conflicts are extrapolated to have lasting impact on environmental outcomes. Such studies fail to take into account the changes in institutional structures over time, which spur the transformation of corporate interests. Issues that initially raised fevered corporate opposition often become nonissues in subsequent decades. The institutional literature provides several insights that help explain the surprising malleability of interests over time. In short, interests are transformed by (1) changes in the regulatory environment, (2) economic dynamics in an organizational field, (3) changes in competitors' calculations, and (4) shifts in consumer preferences.

The conventional literature offers surprisingly little in the way of explicit theoretical explanation of how actors determine their interests (Arrow 1974; Rosecrance 1963). Nevertheless, one image is that corporations straightforwardly estimate the cost of a new environmental regulation and the profits at stake. The greater the costs, the more likely the corporation will engage in actions to avoid implementing the regulation, including lobbying legislators, petitioning for exemptions, or simply continuing operations in contradiction of the regulation.

Even this basic model, however, implies that interest calculations will be rooted in parameters drawn from existing institutional arrangements. As Fligstein (2001) argues, corporate interests and strategies do not form in isolation, but instead reflect their organizational and political environments. Calculations include assumptions about the availability of materials, the preferences of market demand, and the operating costs of competitors. These assumptions in turn depend fundamentally on existing legal requirements, specific economic conditions, and suppositions about broader consumer preferences. As the institutional structures that provide the basis for these calculations change, the calculated value of interests should shift as well. This argument seems broadly non-controversial to a variety of theoretical perspectives. Nevertheless, the basic point that interests change as policy structures change is often overlooked in discussions of the environmental regime.

Regulatory structures are the most obvious institutional forces that shape interests. In particular, corporate interest calculations change depending on whether an environmental policy is merely on the drawing board or has been enacted into law. Corporations may oppose a new policy in the early stages of proposal, when the proposed regulation might be altered or scotched entirely. At this point the question is still "if" – after all, proposed legislation may not be passed or objectionable clauses may be deleted.

Once a new law or regulation is put in place, organizations alter their calculations to include the new requirements as a fixed aspect of the regulatory environment. Even if coercive sanctions are unlikely to be applied, the formation of a new environmental policy shifts corporate calculations from "if" to "when." Thus it is common for a corporation to lobby strenuously against an environmental bill before it is enacted, yet fold it into standard operating procedures once it becomes law (A. Hoffman 2001; Kamieniecki 2006; Porter and van der Linde 2000[1995]). Corporations face many regulatory costs, not only those related to the environment. Once a policy has become law, it becomes simply another factor in the costs of doing business.

Secondly, the institutional parameters of the competitive environment are highly consequential for the interests of corporations. Regardless of the absolute cost imposed by a new regulation, relative costs may not change if the regulation is applied to the entire field including competitors. In some cases early adopters may gain market advantages over late adopters. In others, late adopters may benefit from new technologies that decrease the costs of compliance with environmental mandates. Studies have found that new technologies or procedures sometimes prove to be more cost-effective than the prior polluting ones (Wubben 2000). Regardless, corporate lobbyists tend to be more concerned about creating a level playing field, in which environmental costs are imposed equally, than about the specific cost of a new environmental regulation (Wubben 2000; Porter and van der Linde 2000[1995]; Andersen and Liefferink 1997). Consequently, corporate interest calculations differ when a new environmental policy is applied to the entire economic field compared to initial calculations of the costs to an individual firm.

Related to this, firms that fail to adapt to changing regulatory environments may incur large opportunity costs. Porter and van der Linde (2000[1995]) point to the classic example of Japanese and German car manufacturers that developed more fuel-efficient cars in the 1980s, while the U.S. industry naïvely hoped fuel consumption standards would disappear. "The US car industry eventually realized that it would face extinction if it did not learn to compete through innovation. But clinging to the static mind-set too long cost billions of dollars and many thousands of jobs" (Porter and van der Linde 2000[1995], 50). Decision makers have the option to hold out as long as they want, given the typical weakness of sanctions for environmental noncompliance. However, those that resist may become out-of-step with changing market opportunities and investor and consumer tastes.

Finally, the potential costs of being normatively perceived as an environmental villain can influence the calculation of interests. Polluting corporations may pay a price in bad press or public relations, while "green" companies receive image perks. Oil companies responsible for environmental catastrophes such as Exxon and British Petroleum have sought to undo the damage done to their public image (Fourcade 2011; A. Hoffman 2001). Alternatively, products marketed as "green" may attract consumer dollars, even if ecological benefits are negligible (Yearley 1991). Regardless of whether these "green" and "polluter" images match empirical behavior, the images themselves have weight, which can translate into dollars. Under these conditions, public opinion, media coverage, and social movement activism can have significant effects.

In sum, ordinary institutional analysis can go a long way toward understanding the alteration of interests over time. Interest calculations shift in response to new legislation, changes in the competitive environment, changes in consumer expectations, and other factors. Consequently, as environmental regulations become institutionalized over time, businesses change their behavior in response. Even corporations that initially lobbied against proposed environmental policies are likely to alter their practices once a proposed policy becomes law. In addition, the growing cultural salience of the environment has rewritten the meaning of resistance to global environmental goods, a point that will be developed later in this chapter.

World Society and Cultural Meaning

The key insight of world society theory is, simply, that the international community can be viewed as a society. One feature of a society is the existence of a shared set of cultural understandings that provide the basis for communication and social action. Like any society, the cultural meanings of world society are unlikely to be adopted with perfect universality and uniformity. Nor are conflict and contestation over meanings precluded – indeed, conflict over meaning is a central feature of any society. However, without some coherence and agreement on fundamental cultural meanings, world society could not function.

World society scholars often use the term "taken for granted" to characterize the shared understandings of world society (J. Meyer et al. 1997a). This term has sometimes been taken to imply extreme homogeneity and rigidity. If cultural understandings are completely taken for granted, the implication is that they can never be questioned or changed. Nor is it apparent how a taken-for-granted innovation could be resisted or ignored. This chapter unearths some of the flexibility and nuance within world society theory by articulating how cultural meanings embedded within institutions may change over time.

Cultural Meaning in Institutions

The "thick" conception of culture focuses on the cultural meanings embedded within institutional structures. Institutions are composed of cultural understandings and models that infuse domains of activity with meaning and purpose, as well as organizational structures and agents. One important direction in the empirical literature has been to describe historical shifts in the cultural meanings embedded in global institutions. For instance, Berkovitch (2002) examined the changing cultural meaning of women in world society over the past century from a focus on women as mothers and wives toward a view of women as citizens. Barrett (Barrett and Frank 1999; Barrett and Kurzman 2004) similarly analyzed the shifts in cultural meaning of population growth over the past century. Chabbott (2003) analyzed the evolving meaning of development in global institutions. Cultural sociologists and anthropologists have, in a similar vein, unpacked historical meanings in particular international institutions such as the World Bank or within nation-level institutions (Fourcade 2011; Hall 1989; Carruthers and Babb 2012; Ferguson 1990).

These studies trace cultural meanings by analyzing the discourses or codified rules of key international organizations such as United Nations agencies or various nongovernmental associations. Institutions transcend individual organizations, and so the discourses within any particular organization represent just one indicator of broader institutionalized meanings in society. Indeed, it often proves fruitful to examine the heterogeneity of cultural discourses across organizations, or even triangulate meanings across multiple institutional contexts. For instance, one might contrast the meaning of the environment in pro-environmental organizations against the understandings embedded in organizations of the global development regime.

The "thick" conception emphasizes culture-in-institutions, in contrast with conceptions of culture as ideas and messages floating in the ether or culture as reduced to people's attitudes and values. Free-floating ideas are an aspect of culture – for instance, the "viral" memes of the day or ephemeral popular discourses on current topics. But these free-floating ideas generally fail to generate widespread and durable social change unless they are successfully built into institutional structures. The institutional aspect of culture also differs from the role of culture in structural functionalism and modernization theories, in which culture is conceptualized in terms of attitudes or values internalized by individuals. Instead, the influence of cultural meanings for understanding and predicting social action derives from their "embeddedness" in an institutional context rather than as internalized values within a group of individuals.

The "thick" conception dovetails with scholarship in the sociology of culture, such as the "toolbox" metaphor of Swidler (1986, 2001). In this perspective, the societal toolbox is comprised of collectively institutionalized cultural forms that may be flexibly deployed by agents. In her own work, Swidler emphasizes the discretionary, even strategic, use of culture by individuals. In contrast, world society scholars have focused on the composition of the toolbox: which tools are selected for the toolbox, how those tools develop over time, and why some tools are more widely used than others. This difference is one of emphasis – the forest versus the trees – rather than a fundamentally different understanding of culture.

Nevertheless, world society scholars have relentlessly stressed the overriding importance of institutionalized culture – dominant "tools" in world society – as a parsimonious way to explain widespread patterns of activity. In a similar vein, Chapter 4 characterizes individual and organizational actors as "agents" operating mainly under the aegis of institutionalized structures and meanings. Agents may draw on a range of institutionalized meanings from within the larger cultural toolbox, and may do so with selectivity and discretion. However, flexibility in selecting and applying tools from the cultural toolbox is quite different than theories of unfettered agency. Ultimately, culture is usefully analyzed as a property of societal institutions rather than as one residing within the minds of single individuals. This perspective provides a toehold for subsequent arguments about the dynamics of cultural change and conflict, as discussed next.

Change in Cultural Meanings

A key question remains: how can we explain societal changes in cultural meaning? Common alternatives view culture as the product of hegemonic or material interests, or emphasize the agency of key actors or movements who articulate new cultural frames that suit their agendas. World society theory shifts attention away from actors and toward the institutional structures in which they are embedded (J. Meyer and Jepperson 2000). Cultural change is propelled by institutions themselves, which interact and combine to yield new cultural meanings and new institutional forms. The notion that social change originates from macro-social institutions has deep theoretical roots in sociology, but may seem counterintuitive given the prominence of actor-centric explanations in contemporary social science (J. Meyer and Jepperson 2000). Nevertheless, it makes sense to trace the process back to macro-social institutions, which provide the cultural building blocks and collective purposes that yield new cultural forms. Agents mediate this process of change by interpreting and combining meanings from existing institutions, and improvising in pragmatic and creative ways (Hallett and Ventresca 2006; Dobbin 2009; Boyle 2002; Fligstein 2001). While agents play an important mediating role, cultural change is not reducible to agents or their material or to bureaucratic interests. Instead, cultural change is best understood as originating from institutions themselves.

This institution-centric approach to culture resembles Foucault's approach to historical change. Discursive formations or *epistemes* emerge from a constellation of changes in prior discursive domains. For instance, Foucault argued that new discourses regarding sexuality emerged from the interplay between previously institutionalized religious discourses, emergent medical/scientific discourses, and educational institutions (Foucault 1990). The process involved individual agents – priests, doctors, psychoanalysts, and educators – but could be more fruitfully analyzed in holistic terms as the interaction of discourses, rather than the accomplishment of particular individual or organizational actors.

A similar explanation can be developed for the emergence of new cultural meanings of the environment from prior institutionalized meanings and structures. For instance, Chapter 2 examined how new post-Stockholm environmental discourses and meanings emerged out of the interaction and recombination of cultural frames drawn from the international development regime, and earlier pro-environmental treaties and institutions. Similarly, Chapter 3 argued that experts working within the institutional context of science deployed scientific frames in conjunction with

environmental frames, ultimately reshaping the meaning of ozone depletion across the global environmental regime. Agents reshaped the cultural
meaning of hazardous waste by extending preexisting institutionalized
discourses relating to health and medicine, as discussed in Chapter 4.
These processes of change are inherently contingent – one cannot analyze
future cultural meanings as a simple extrapolation of prior institutional
meanings. To further complicate matters, the influences of preexisting
institutional structures on emergent meaning are not necessarily unidirectional – institutions are likely to mutually affect each other. The international development regime clearly influenced the nascent Stockholm
environmental regime. However, the global environmental regime has
subsequently reshaped institutionalized understandings of development –
reflected in the emergence of contemporary discourses of "sustainable"
development for example.

Cultural change involves processes of agglomeration, recombination,
and syncretism, in which discourses and meanings are grafted or incorporated from other institutions. As institutions intersect over time, the
cultural meaning of the "environment" takes on new layers of meaning.
While the current global environmental regime was built largely from
building blocks of earlier pro-environmental institutions, one can also
observe the residue of several other global institutions. The environmental regime is multifaceted, having incorporated logics and discourses
from international institutions and regimes related to health, economic
development, science, and others.

Just as new layers of meaning become affixed to institutions, earlier
meanings and connotations may be effaced. In the 1950s, smokestacks
with billowing clouds of smoke were meaningful as indicators of progress and economic prosperity (Brenton 1994). This meaning has been
displaced in the global North as additional meanings have accumu-
lated over time, including connotations of environmental degradation as
well as negative health effects. "Industrialization" itself has developed
a retrograde connotation – a relic of the past rather than a symbol of
the future. Although the objective appearance of the smokestacks may
remain unchanged, their meaning has substantially altered over the past
several decades.

The cultural meaning of the environment is flexible but not infinitely
malleable. Formal institutional structures provide an anchor that stabi-
lizes meanings over time, at least in the medium term. Simultaneously,
however, some cultural meanings or discourses fade while new meanings emerge and become more central in global discourses. A slight

twist in meaning may be surprising or non-intuitive the first time it is encountered, but it becomes commonplace and taken for granted fairly rapidly. Consequently, cultural meanings can change dramatically over decades, even though change may be hardly perceptible from year to year. Swimming in the sea of cultural meaning, it is difficult to recognize the accumulation of new connotations without conscious effort. Thus we are carried along in an ever-changing context of cultural meanings, shaped by the interactions of the broader institutions in which cultural meanings are embedded. It is only through historical comparison to prior decades or centuries that the evolution of cultural meaning becomes obvious – and even then, one must be vigilant to avoid imbuing contemporary frames and meanings into the interpretation of past discourses and activities.

Meanings in World Society: Beyond the Cultural Juggernaut
This chapter emphasizes the contingent nature of cultural change. The implicit alternatives include classical modernization theory, and various caricatures of world society theory, which explain contemporary institutions as the teleological culmination of inevitable historical trends. World society scholars often discuss sweeping historical changes, such as the rise of individualism or scientific rationalization. This focus on macro-cultural change can be mistaken for "cultural juggernaut" arguments, in which potent discourses or institutions (often traced to the Enlightenment) steamroll forward across centuries, leaving a culturally homogeneous world in its wake. By contrast, this chapter stresses the contingency involved in cultural change. This approach suggests mapping the accretion of incremental change in the meaning of the environment as a consequence of interactions between the global environmental regime and other global institutions relating to development, health, human rights, and so on. Like Foucault's historical analyses of discursive change, such research cannot make claims of unidirectional causality but instead might chart the agglomeration of connotations that gradually shift cultural meanings across decades. This section unpacks three issues or assumptions underlying this approach to the analysis of culture.

First, the institutional conception of culture departs from traditional structural-functional analysis in that there are no presumptions of overall coherence: there is no intrinsic logic to institutions, nor do they fulfill any necessary role within society. Scholars may examine or document secular cultural trends – for instance toward greater individualism or human rights – but the approach is ultimately empirical and historical rather than teleological or normative. The emergence of a human rights regime is something to be explained, not the inevitable consequence of societal modernization or the inexorable extension of ideologies born in the Enlightenment. Institutions are fundamentally rooted in cultural meanings, but the particular content of these meanings evolves in path-dependent ways through contingent events and interactions with other institutions. Moreover, institutions may not be wholly logically coherent or internally consistent. Cultural and organizational fields are complex, and loose coupling is commonplace. Consequently, subsequent cultural meanings and formulations cannot be neatly extrapolated from the logic of prior institutional structures and meanings.

Second, institutions are not prepackaged with sharp, fixed boundaries. Rather, cultural meanings are embedded in a set of potentially overlapping institutions, which may evolve independently and jostle for jurisdiction over time (Fligstein 1997). Consequently, cultural meanings may vary across different institutional contexts. For instance, understandings of the environment in UNEP may be subtly different from those in the World Bank or the global human rights regime. Moreover, the primary locus of embeddedness may shift across institutions over time, transforming taken-for-granted meanings in the process.

Finally, this institutional conception of culture takes no stance on the normative status or objective value of cultural meanings in a given historical era. Cultural orientations toward human rights or environmental protection are not reflective of universal truths or the inevitable product of moral imperatives. Rather, they are distinctive historical constructions that might have evolved quite differently. This relativistic view of culture is commonly adopted by anthropologists but may be more difficult for environmental protesters in the barricades to accept. The landscape of cultural meanings shifts continually, and there is no place to stand "outside" of our culture to judge whether the direction of movement is objectively for the better. This perspective encourages agnosticism about whether objective progress is being made on the environmental front. Subsequent generations, operating under a future set of cultural meanings, may look upon contemporary environmentalism as odd or misguided. Or, if environmental institutions and meanings remain fairly stable, the contemporary period may be perceived in hindsight as a watershed. At best, evaluations can be made about perceptions of progress according to the cultural meanings of the day.

In sum, cultural meanings are a core component of institutions. However, their fundamental, taken-for-granted nature makes analysis difficult. Historical changes in cultural meanings are of scholarly interest in themselves, and they are essential for understanding large-scale historical changes in social action, as discussed in the next section.

Cultural Meanings and Social Change

Cultural meaning plays a fundamental role in the Bee Swarm model of social change. Culture is less tangible than formal policies and vocal agents, and consequently, institutional studies often focus on the latter two. However, culture plays a critical role at every stage: New meanings are central to the social construction of social problems, rendering new issues actionable (see Chapter 2). Cultural meanings and frames motivate the construction of new institutional structures and give purpose to new institutional agents – animating and propelling the Bee Swarm (Chapters 3 and 4). Moreover, cultural change transforms the interests of actors within a cultural field, as previously addressed. Finally, long-term shifts in cultural meaning render older cultural frames as antiquated, deviant, or incomprehensible – ultimately undermining even the staunchest resistors to social change. This process is taken up here.

World society theory claims its theoretical foundation in the social construction of reality (Berger and Luckmann 1980[1967]; Schutz 1967). While objective physical reality exists independently of humans, social interpretation is required before humans can act on this reality. In the environmental arena, physical realities must be constructed as meaningful social problems before people can address them (Gusfield 1981). Global institutions serve to stabilize and propagate particular meanings, often leading to the diffusion of new institutional structures from the centers of world society to the periphery. At times, the new meanings wholly supplant prior meanings, rendering those who have stubbornly resisted change as outmoded or anachronistic – their actions cease to make sense within the contemporary landscape of meanings.

One might imagine that a similar process occurs in language. A word, such as "texting" emerges, giving meaning to an activity previously described clumsily as "sending messages by phone." This new meaning is initially limited to innovators in cell phone usage but may spread through society, filtering down to those who are peripheral. Challenging or questioning the new usage is easy in the early days. As a new term becomes widely institutionalized, appearing in dictionaries and broadly in societal discourses, it becomes much harder to resist or challenge the usage. Over time, the remaining resistors are likely to seem antiquated and irrelevant.

Eventually only a handful, perceived as technological luddites or otherwise peripheral individuals, reject the new terminology.

Institutionalization does not necessarily generate complete isomorphism, or preclude resistance to institutionalized cultural meanings. Indeed, the establishment of new or expanded institutional meanings often prompts strong reaction. Resistance is a consequence of institutionalization, not just a potential obstacle to it. However, the process of institutionalization, by changing accepted cultural meanings, may erode the justifications for resistance and eventually renders resistance archaic and outmoded.

Conflict and the Emergence of New Meanings

According to world society theory, certain forms of conflict ought to decrease as the global environmental regime becomes institutionalized. In particular, contestation over the broad legitimacy of the environment on the global political agenda should decline as institutionalization occurs. Environmental values have been added to the panoply of sacred societal goals to the extent that it is awkward to publicly declare oneself proudly as a polluting state or an environmentally ignorant corporation, as was acceptable in previous decades. However, the cultural meaning of the environment changes constantly over time. Consequently, conflicts are likely as the environmental institution expands its agenda, encroaching on the cultural meanings embedded in other institutions. The conventional perspective assumes these conflicts signal the decreased ability of environmental institutions to alter practices in key areas such as climate change or biodiversity. Paradoxically, however, these conflicts between institutions can be seen as a positive indicator of institutional strength.

One common type of conflict over cultural meanings occurs along institutional boundaries. As an institution seeks to expand the slate of cultural meanings within its jurisdiction, it may conflict with cultural meanings embedded in neighboring institutions. Conflict may ensue as the cognitive boundaries between institutions are brought into question. Institutional expansion does not necessarily entail such conflicts – in most cases the cultural meanings of multiple institutions agglomerate without fuss. However, when such conflicts do occur, they can be seen as an indicator of the expansionary strength of an institution rather than its weakness.

Since its inception, the global environmental regime has conflicted over cultural meanings with other institutions devoted to health, economic development, and national sovereignty, among others. One instance of this was discussed in Chapter 2, as the initial formation of an international environmental regime jostled its way into the existing field of global institutions. As the global environmental regime has developed, the expansion of its agenda has resulted in proposals for environmental standards in a growing set of domains. For instance, the breadth of the proposed environmental protection actions regarding climate change has created a broad array of institutional opponents, as discussed later in this chapter. These conflicts need not read as the result of decreased effectiveness of the environmental regime. Instead, it is the strength of the environmental regime that has enabled climate change mitigation proposals that could dramatically affect the workings of the economic institutions of society.

This reinterpretation of institutional boundaries due to conflict or incorporation with aspects of other institutions provides a major means of alteration in the cultural meaning of the environment. Conflicts with other institutions shift the cultural meanings on both sides, changing dynamically with the continuous alteration of environmental meaning in society. The outcomes of these conflicts within the global environmental regime should not be understood as simple winning or losing. Instead, the outcome of these conflicts results in shifts in frames and boundaries for both institutions, rather than one institution simply winning without accompanying alteration.

Conflict between institutions is often perceived as a cause of failure or at least a setback for environmental progress. However, from a macro-level historical perspective, conflict can be understood as resulting from the expansionary strength of an institution. An institution that is pushing its boundaries by encroaching on the issue jurisdictions of other institutions is likely to manifest conflict; a moribund institution will be more passive as it seeks merely to maintain its existing boundaries. Conflicts between institutions do not necessarily signal the failure of a particular agenda item in the long term. Regardless of any particular outcome, institutional conflict may broadly be seen as a sign of the energy of an institution as it expands structurally, recruits more environmental agents, and extends its boundaries into new domains.

Diffusion and Adoption of New Meanings

World society scholars have shown that organizational forms and policies tend to diffuse in a global pattern (J. Meyer and Hannan 1979; Suchman and Eyre 1992). This diffusion can be understood as the growing salience of a new set of cultural meanings. Diffusion follows a typical pattern:

adoption occurs first among those most central to global "conversations" and spreads outwards to those less engaged in global institutions and discourses.

This diffusion pattern can be explained by recognizing the importance of shared cultural meanings in a society. New cultural meanings make most sense to those deeply embedded in world society. Since Western nation-states tend to be centrally involved in global discourses, they tend to be early adopters in the diffusion process. This adoption occurs even though the actual costs are usually higher for core nations than for those in the periphery. In the case of environmental protection in particular, new meanings typically promise much higher costs and inconvenience in industrialized countries. Nevertheless, those countries – which are most central to the discourses in world society – have tended to most readily adopt the cultural meanings embedded in global environmental institutions.

As these culturally central states increasingly begin to adopt the new cultural meanings, the meaningfulness of the cultural frameworks expands in world society. As the use of a new word gathers momentum, it spreads beyond the initial group that invented it; analogously, those states that are semi-peripheral to the global cultural discourses begin to adopt the new cultural meaning as well. The cultural meaning might not correspond as neatly to domestic conditions or discourses for those states more removed from global discourses, but the broader meaningfulness of the concept in world society encourages adoption. An occasional cell phone user may pick up trendy technological talk just from participating in general social discourses. Similarly, culturally semi-peripheral nation-states may adopt global cultural meanings through occasional participation in global discourses.

Finally, states that are peripheral to the world society may eventually adopt the new cultural meaning. For those states, their general lack of participation in global discourses makes them less aware and less eager to adopt new cultural meanings. Domestic cultural meanings in peripheral states may map less well onto world society discourses, providing little incentive for adoption. Eventually a cultural meaning may become so widespread and "taken for granted" in world society that even peripheral nation-states adopt the new meaning. In other cases, cultural meanings may never diffuse to the distant periphery. This lack of adoption does not signal the active resistance of peripheral nation-states but rather a lack of engagement with global conversations.

Scholars have tended to focus on the mechanical organizational processes of diffusion, rather than the cultural meanings involved. This organizational focus has obscured the subtleties of the alteration in cultural meaning. Cultural meanings do not flow in only one direction. Instead, interactions with various domestic institutions as well as with other global institutions alter cultural meanings locally – and sometimes even globally – as they diffuse throughout world society. Future research would benefit from more nuanced attention to the influence that meanings in semi-peripheral and peripheral nation-states have on world society. Such an analysis would not assume unilateral influence of core on periphery, but on the mutual give and take of institutional interactions.

Resistance, Anachronism, and Social Change

The Bee Swarm model seeks to explain social change in a situation where global institutions are easily resisted, as is the case in the global environmental regime. The cost of a single bee sting is nominal, such as mild censure by international organizations or an embarrassing protest by an NGO. Resistance – the failure of states or firms to adopt new environmental regulations – entails few direct economic or political costs. Instead, the Bee Swarm model focuses on the accumulation of small pressures and costs. An attack by a large swarm of bees can be deadly.

Changes in cultural meaning play an important role in the Bee Swarm, compounding the effects of structures and agents described in previous chapters. Changes in institutionalized cultural meaning bring a host of new pressures, expectations, and costs for resisters, as discussed in Chapters 3 and 4. However, cultural change has another consequence that is less often discussed: Those that resist changes in cultural meaning over the long term ultimately risk becoming an anachronism. As cultural meanings and debates move forward, those stuck in the positions of the past increasingly become marginalized – not merely stigmatized, but rendered incomprehensible and irrelevant. These cultural costs, which are ultimately the costs of reduced participation in world society, can be sufficient to motivate change even in die-hard resisters.

One might draw an analogy with changing fashion trends. Those most central to the cultural field of fashion will most quickly adopt the latest styles, while those on the periphery will adopt slowly, if ever. Individuals may resist trends for many reasons, such as the economic savings that result from infrequent replacement of one's wardrobe. However, being stuck in yesterday's clothing does not prevent the fashion world from moving forward. As one falls behind, the dowdy may be penalized on social occasions such as job interviews or first dates. Akin to a bee sting, the cost of any one penalty may be minor. Over a decade or two, however, the failure to keep up with fashion trends begins to mark oddity rather than individuality. What might have begun as a personal stance or economic choice regarding fashion ultimately becomes stigmatized or incomprehensible to onlookers, or is seen as disengagement from society. Few remain immune to changing meanings within their social group, since immunity equates with ostracizing oneself.

In short, the resistance of a handful of individuals or states – even powerful ones – usually does not prevent change in world society. In the environmental realm for instance, cultural meanings have changed greatly in the corporate world over the past several decades. Andrew Hoffman (2001) has found that in the 1960s and 1970s, corporations generally sought to prevent the governmental imposition of environmental regulations. However, by the 1990s, corporations had shifted their interest calculations to proactively implement environmental behaviors into their operating procedures. As Hoffman (2000, 127) summarizes, "the 'rules of the game' have changed. Managers acting in the best interests of their investors must now consider environmental protection in their decision making." Although some corporations continue to resist these changes in meaning, resistance has not prevented broad cultural change in the field as a whole.

Second, the development of new cultural meanings permits and even encourages a variety of sanctions and penalties against the resistant organization. Observers may mistake these sanctions as the main levers of change, without recognizing the degree to which the mechanisms are embedded within broader cultural meanings. The resistant corporation becomes increasingly pressured by a diversity of organizations toward more current cultural meanings of the environment. Shifts in cultural meanings broadly affect competitors, insurance companies, financial institutions, the media, the local community, consultants, and activists, regardless of the resistance of any particular corporation (A. Hoffman 2001, 15). As Van den Akker (2000, 134) notes, "banks, accountants, insurance companies and – last but not least – consumers are becoming more and more demanding and such demands include environmental requirements." As Hoffman concludes, "To risk contradicting the accepted norms of the day could force social censure in one of many possible forms – legal penalties, public protests, inability to gain liability coverage, and so on" (A. Hoffman 2001, 14). While an organization may still resist change, strongly institutionalized meanings encourage numerous bee stings for the resistant organization.

Finally, the resistant firms and states risk the intangible danger of anachronism or irrelevance. In a world that is continually moving forward, an organization still clinging to the practices of the past becomes increasingly perceived as old-fashioned. Corporations that ignore environmental issues increasingly risk the appearance of being behind the times in a corporate world that is chasing the next new thing. Nation-states face similar pressures within world society. In the 1970s, for example, it was culturally meaningful for states in the global South to proclaim their desire for more industrial pollution, drawing on the cultural idea that pollution indicated economic progress. Brazil was a particularly fervent anti-environmental state. In the 1980s, Brazil opened its borders for hazardous waste disposal and lobbied against international efforts at forming a substantive forest policy (Hecht and Cockburn 1989). However, as global environmental institutions continued to shift the cultural meanings of toxic waste and deforestation, Brazil's position appeared increasingly shortsighted, odd, and even backward. Subsequently, the regime of President Collor de Mello implemented policies consistent with up-to-date environmental meanings in the 1990s (Hurrell 1992, 398).

Resistance to mild norms is easy to imagine, given the weight of economic interests coupled with the trivial sanctions imposed for breaching environmental norms. However, resistance to broad cultural shifts in the meaning of the environment in the long run is akin to disengagement from society. As new cultural meanings are adopted throughout society, resistant corporations and nation-states are seen as increasingly deviant. Such cost is insignificant for those already peripheral, but may be serious for those central to world society. The punishment for long-term resistance of these cultural meanings is not a rap on the knuckles, but the risk of obsolescence, fighting a battle that has already been lost.

These institutional dynamics are particularly clear in the case of climate change discussed in the following section. From the "thin" normative perspective, the United States appears to have (thus far) successfully resisted world society pressures for the adoption of climate change policies. However, from the "thick" cultural perspective, the resistance of the United States has not prevented the expanding cultural meaningfulness of climate change in the global community and even within the United States. Instead, numerous subnational states, cities, and corporations within the United States have developed climate change policies, offering a patchwork of policies at the national level. As the issue of climate change becomes increasingly institutionalized in the global community, continued U.S. resistance may eventually be perceived as puzzling or self-defeating rather than as a rational defense of economic interests.

The Case of Climate Change

At first glance, the effort to create an international regime on climate change appears an ideal case to argue against the world society perspective, because corporate interests have successfully blocked efforts to build a strong global regulatory regime. By 1992, pro-industry interests had largely defeated early efforts to address the problem of climate change. However, in the years since 1992, pro-environmental agents in world society have continued to advocate for a stronger international climate change regime. Victory is far from certain, and current efforts have not yet produced major reductions in greenhouse gases. However, just as the physical process of climate change is a slow process with a great deal of momentum, the continued Bee Swarm of international efforts may be cumulating over the long term.

The first major push for the formation of an international policy structure controlling greenhouse gases was the creation of the UN Framework Convention on Climate Change (UNFCCC) in 1992. The treaty, heralded at the 1992 UN Conference on Environment and Development in Rio de Janeiro, was the culmination of four years of preparation behind the scenes. With great flourish, the treaty was signed by 159 states, putting greenhouse gas emissions prominently on the international agenda. However, the actual specifications of the treaty were viewed as weak and disappointing. One scholar notes with lukewarm enthusiasm that "The general assessment has been that while, from an environmental point of view, it is clearly inadequate, it is possibly as good a political compromise as could be reached given the constraints" (Paterson 1996, 64).

Nevertheless, the UNFCCC has led to the expansion of public awareness of the issue of climate change. Popular jokes began to abound, such as the improvements that global warming might make in Minnesota or the creation of Lake Los Angeles. More seriously, major U.S. magazines such as Time and Newsweek began to regularly cover aspects of climate change (Kluger 2006; Begley 2011). Energy efficiency, in products ranging from light bulbs to hybrid cars, touted in the 1970s as an answer to diminishing oil resources, is now portrayed as addressing climate change. Today, it is difficult to be wholly unaware of the issue of climate change, although continuing conflict and resistance over its meaning (including its ontological status) leads to a range of positions. These conflicts are arguably a testament to the strength of the global environmental regime to challenge the boundaries of the powerful and deeply institutionalized global capitalist regime.

The 1992 UNFCCC and the subsequent Kyoto Protocol of 1997 have certainly not had much effect on the world's output of greenhouse gases. Drawing on the "Smoking Gun" model, the regime is largely ineffective. However, the new cultural meaning of climate change has set in motion a wide range of actions aimed at creating more structures, engaging the efforts of environmental agents, and expanding the cultural meaningfulness of climate change through educational efforts. The jury is still out on whether these Bee Swarm effects will cumulate to the point that they ultimately reduce the damaging effects of human activity on the planetary atmosphere. However, in the twenty years since the initial formulation of the UNFCCC and the Kyoto Protocol, there have been substantial shifts in the meaning and salience of climate change that promise improvements in the future.

Conflict Results from the Expanding Environmental Agenda

The tepidity of the Kyoto Protocol of 1997 has been widely attributed to the opposition of corporate interests. Firms in both the United States and Europe banded together during the preparations for Rio in 1992 to lobby against the creation of the Kyoto Protocol to an extent unprecedented in previous international environmental negotiations (Fisher, 2004; Harris 2000; Selin and VanDeveer, 2009). The formation of an unusually broad and powerful corporate lobby reflected the breadth of proposed changes for businesses in industrialized countries, as well as the extent of the reductions theorized as necessary for climate change amelioration. Viewed from the world society perspective, the degree of opposition to global climate change proposals can be seen as an indicator of the temerity of the proposed boundary expansion of the global environmental regime.

The global environmental regime previously proved its mettle in successful efforts to limit the production of gases such as sulfur dioxide (acid rain) and chlorofluorocarbons (ozone depletion). The regime is seeking to expand its institutional boundaries even further with the regulation of greenhouse gases. The consequences of global warming, including alterations of temperature and precipitation, sea level, and consequent effects on animal and plant life, are predicted to be of alarming scope (IPCC 2007, NRC 2010). Nevertheless, the attempt to mandate decreased production of greenhouse gases is a mind-blowing undertaking. Greenhouse gases are a barometer of industrialization itself, at least as currently practiced, since these gases are produced largely by the consumption of the fossil fuels that provide the backbone for industrial economies. It is not surprising that this effort to so greatly expand the institutional boundaries of the global environmental regime has generated significant conflict.

Previous targets of the global environment regime were more limited, as in the cases of chlorofluorocarbons and sulfur dioxide. Although environmentalists had despaired at the time over the difficulty of getting powerful industrial corporations to alter production of these substances (as partially discussed in Chapter 3), in retrospect, the production sites for chlorofluorocarbons and sulfur dioxide seemed small when compared to the point sources for greenhouse gases (Benedick 1998; Soroos 1997). Nearly every business sector in the industrialized nations would be affected by proposed changes in greenhouse gas emissions. Institutional conflict is further aroused because transnational corporations are estimated to produce half of all greenhouse gas emissions in the world (Paterson 1996, 160). As currently conceptualized, climate change issues affect "virtually every major economic and social function … from transportation to agriculture, and from land use changes to industrial processes" (Falkner 2008, 97).

Not only is the issue jaw-dropping in scope, the size of needed greenhouse gas reductions is staggering. Climate scientists have theorized that dramatic reductions in greenhouse gas emissions will be required in order to improve global warming within a reasonable time frame. The Kyoto Protocol requires a significant decrease in emissions for major producers of greenhouse gases such as the European Union and the United States, although even the proposed decreases are thought to be insufficient. As spokespeople for the fossil fuel industries have argued, "action to reduce emission of greenhouse gases, on anything like the scale suggested by the majority of climate scientists, would be highly undesirable given its potential to affect fundamentally the way in which they currently operate" (Newell 2000, 97). Even if such drastic actions are necessary for environmental sustainability, such reductions can only be attained with major cost and effort in nearly all walks of industrialized life.

As one might expect, substantial opposition has been raised against the possibility of worldwide standards for greenhouse gas emissions. The anti-environmental Global Climate Coalition, which was formed in the initial years of preparation for the UNFCCC, was composed of over 40 corporations and was dominated by fossil fuel industry interests (Harris 2000). Scholars have attributed the lack of binding targets and timetables in the Kyoto Protocol in large part to the aggressive lobbying of the Global Climate Coalition and other business groups (Harris 2000). Consequently, the Kyoto Protocol has been viewed at best as a mild first step in the efforts to curb greenhouse gas emissions. From the point of view of environmentalists, corporate interests scored a major victory in the first round.

From the standpoint of world society theory, however, the climate change issue can be seen as a remarkable effort by the global environmental regime to extend its jurisdiction into the heart of capitalism. It is unsurprising that this audacious expansion of the global environmental regime would lead to serious conflict with other institutions – in this case, industrial capitalism. This is not to suggest that the climate change issue is a fiction designed merely to allow the symbolic flexing of institutional boundaries – scientific evidence persuasively argues that greenhouse warming is an urgent and real problem facing the planet (IPCC 2007, NRC 2010). Nor does it assume that the global environmental regime will necessarily be successful in this impudent attempt to extend its boundaries. However, it is difficult to imagine the global environmental regime mounting such a challenge back in 1972, when it had barely managed to get established amid the competition of its United Nations sibling agencies. The ability of the global environmental regime to mount such a presumptuous agenda suggests astonishing institutional strength rather than weakness.

The Re-formation of Corporate Interests

In the early 1990s, corporations in both the United States and Europe viewed international efforts at regulation of greenhouse gases as obviously contrary to their economic interests. Business interests on both sides of the Atlantic joined the Global Climate Coalition and mobilized to block international efforts toward a climate change regime (Newell 2000, 104). However, the different institutional structures in the United States and the European Union have encouraged different calculations in the aftermath of the UNFCCC. In the United States, corporations figured that domestic institutional structures might never change – calculating "if" rather than "when" – which encouraged corporations to maintain their pre-Kyoto interests. In contrast, European corporations operating under different institutional conditions saw that it was only a matter of time until new regulations curbing greenhouse gas emissions would be enacted. Although the actual provisions of the Kyoto Protocol were mild, European corporations foresaw more substantive and binding regulations as inevitably on the horizon. Consequently, their calculations shifted toward finding advantage in the new regulatory climate.

The United States was an early leader on the issue of climate change in the 1980s (Pring 2001). However, the U.S. presidential regimes during the 1990s took an oppositional stance to climate change proposals. Throughout the negotiations for the UNFCCC, the United States obstinately refused to consider specific limits on greenhouse gas emissions. The weakness of the resulting Kyoto Protocol has been widely attributed to the intransigence of the U.S. position, which closely represented the interests of U.S. and European industrial corporations (Harris 2000). The subsequent refusal of the United States to ratify the Kyoto Protocol additionally weakened the effectiveness of the international greenhouse gas regime. As one scholar summarizes, "It seemed that the obstructionist stance of the US fossil fuel industry had paid off" (Falkner 2008, 128).

In the United States, the lack of change in government institutional structures on climate change issues has supported the maintenance of hard-line anti-climate change interests. The presidential administration of George W. Bush was particularly unsympathetic to climate change policies, as were the Republican-dominated Congresses (Newell 2000, 163; Paterson 1996, 87). Under these conditions, anti-environmental corporations calculated that climate change regulations might never be instituted in the United States. With no change apparent on the horizon, anti-environmental corporate interests in the United States had little reason to shift. As one scholar notes, "For many US firms, the Kyoto Protocol did not become an immediate reference point as soon as it was adopted in 1997, but continued to be a contested and potentially irrelevant form of international regulation that might never enter into force" (Falkner 2008, 125).

In contrast, anti-environmental interests in Europe faced a different institutional environment. Although European corporations were equally opposed initially to limitations on greenhouse gas emissions as their American counterparts, European domestic and regional governments were not nearly as accommodating to their interests. Although greenhouse gas reductions would be as costly in Europe as in the United States, it became evident to European corporations that the ratification of the Kyoto Protocol by the European Union meant the eventual implementation of greenhouse gas regulations. Even during the UNFCCC negotiations, some European corporations shifted their positions and offered to cooperate with government implementation. As Falkner (2008, 125–126) writes, "If anything, the political salience of climate change continued to rise in Europe and businesses could therefore realistically expect the protocol to be ratified across Europe, or would at least see Kyoto-type measures enacted."

After the Kyoto Protocol was signed, the different institutional structure in Europe led to recalculation of interests by European corporations.
Although the Kyoto Protocol had set only weak targets, the creation of
an international treaty on climate change was understood by European
corporations as setting "a precedent for a future tightening of international commitments ... and there was no guarantee that the fossil fuel
lobby could maintain a united business front in a changing political environment" (Falkner 2010, 107). Even before formal ratification, European
corporations calculated that, "international restrictions on carbon-based
energy sources were no longer a remote possibility but an increasingly
realistic scenario. ... Businesses began to factor in the costs of climate
action and demanded a stable regulatory environment for climate policy"
(Falkner 2008, 124). As another scholar comments, "Because of the state
of play in the climate debate in the US, [U.S. corporations] have been
able to centre on whether or not climate change is a problem at all. In
Europe the debate has moved on to which response is more appropriate."
(Newell 2000, 121)

In short, the different institutional conditions fostered recalcula-
tions of corporate interests in Europe, leading to fractures within the
international corporation coalition. Even within the hard-line fossil
fuel lobby, differences began to emerge between gas and coal interests
(Newell 2000). The leading European oil firms of British Petroleum and
Royal Dutch/Shell both left the oppositional Global Climate Coalition
and have chosen instead to take advantage of investment opportunities
in friendlier climate change technologies such as renewable energy and
photovoltaics (Newell 2000, 120). Shell has already become one of the
major players in the solar energy industry (Wubben 2000, 21). In 1996,
the American subsidiary of British Petroleum also withdrew from the
Global Climate Coalition, "in a move that signaled the deepest rift yet
within the fossil fuel sector" (Falkner 2010, 108). Other defectors from
the Global Climate Coalition have included the Ford Motor Company,
DaimlerChrysler, General Motors, and Texaco by 2000 (Pew 2006).

While many corporations in the U.S. energy industry have held onto
the hope that greenhouse gas emissions will never be regulated, many
other U.S. corporate sectors have begun to see new economic opportunities that might arise out of the proposed massive economic shift. Major
companies such as Toyota, General Motors and Philips have "moved
from a defensive toward a more pro-active and offensive position," recognizing "the need to adapt to the public opinion at large and ... major
market opportunities" (Wubben 2000, 21). One of the leading sectors

in the pro-environmental lobby has been the global insurance industry, led by corporations such as Munich Re and Swiss Re, which has realized that extreme weather patterns are likely to lead to worldwide increases in insurance costs. Other corporate sectors including banking, sustainable energy, gas, insurance, and environmental technology have also begun to grasp the opportunities that might arise from a more stringent climate change regime (Wubben 2000, 21).

Since the 1992 UNFCCC, the landscape of corporate interests has undergone substantial reformulation, in line with the different institutional promises of various governments. This has led to the formation of moderately pro-environmental business interest coalitions in international climate change forums, such as the International Climate Change Partnership (ICCP), which includes important chemical and electronics manufacturing firms such as AT&T, Electrolux, Enron, General Electric and 3M (Pew 2006). International business no longer presents a monolithic front opposed to climate change proposals, but has splintered into a variety of coalitions, enabling negotiations to "draw more moderate voices into a constructive dialogue on how to reduce the technical and economic costs of climate action." (Falkner 2010, 109). As climate change policies have become an institutional reality in many countries, corporations have accordingly shifted their calculations and their interests.

Business interests reflect the institutional structures that underlie the calculation of those interests. When the institutional structure can be assumed to be constant, corporations face little motivation to shift their interests, as illustrated by the intransigence of many major corporations in the United States. However, when institutional structures change, savvy corporations recalculate their interests in order to take maximum advantage of the opportunities offered by the regulations as well as to minimize the costs. This accords with the world society expectation that shifts in institutional structures, such as the negotiation of international treaties and the passage of national legislation, will eventually encourage the transformation of interests.

Resistance to New Cultural Meanings

Perhaps the most visible obstacle to the success of a global climate change regime thus far has been the antagonism of the U.S. government, largely prompted by the lobbying by U.S. corporate interests. However, despite the formal resistance of the U.S. government, the "thick" meaning of climate change issues has begun to shift both within world culture and within the United States itself. Changes are occurring rapidly within the United States itself, as the cultural meaningfulness of climate change has shifted, influencing various organizations including U.S. states, cities, and corporations. Perhaps most visibly, the issue of climate change has dramatically increased in salience over the past few decades, from an esoteric issue of planetary scientists to a highlighted topic of political controversy and general conversation. However, the outcome of global efforts to reduce climate change is still undetermined.

The United States has appeared to stand like a rock resisting the tide toward greenhouse gas regulation. Will the rock win? Since the United States is the world's second largest producer of greenhouse gases – China is the largest – its position on climate change has had a significant impact on the effectiveness of an international climate change regime. Scholars have argued that the U.S. refusal to ratify the Kyoto Protocol has decreased the normative value of the treaty. Indeed, Jacques et al. (2008) have argued that the position of the United States has influenced other developing countries such as India and China, which have also refused to ratify the treaty. In other words, resistance by the United States has weakened the force of international environmental norms.

Thus at first glance, it seemed that U.S. corporate interests succeeded in making the United States a haven for greenhouse gas producers. However, the complexity of the institutional field may question that initial assessment. Given the immobility of the U.S. government, subnational governments within the United States have begun taking their own action on climate change. The lack of U.S. federal legislation on climate change has created a policy vacuum that is being eagerly filled by state and local governments. Various constellations of subnational actors – including U.S. states, Canadian provinces, municipalities, universities, and corporations – have voluntarily pledged to reduce greenhouse gas emissions, enact regulations, and promote new practices to reduce greenhouse emissions. By June 2011, the U.S. Conference of Mayors Climate Protection Agreement had been endorsed by 1,053 cities, ranging from New York City, to smaller American cities with less than 100,000 inhabitants, such as Kenosha, Wisconsin. The mayors who have signed on to the agreement come from all 50 of the United States and represent a total population of over 88 million American citizens (Fisher 2013). In 1993, an international coalition of cities, the Cities for Climate Protection (CCP), arose in which municipal governments in different countries undertook to alter policies on land use, transportation, and energy management in

order to curb greenhouse gas emissions. By 2006, worldwide membership included 674 cities including many in the United States (Lutsey and
Sperling 2008).

U.S. states have also responded to the threat of global warming within
their political jurisdictions, joining climate change initiatives and creating inventories of greenhouse gas emissions (Fisher, Waggle, and Leifeld
2013). Some states have also enacted laws that set emissions targets for
greenhouse gases. For instance, in 2006 California pledged to reduce
greenhouse gas emissions by 25 percent by 2020 and sought to apply
controls to utilities, refineries, and manufacturing plants (California
Energy Commission 2006). Regional compacts have also formed among
U.S. states in the West and Northeast, such as the Regional Greenhouse
Gas Initiative and the New England Climate Coalition (Krane 2007,
463). As one scholar notes, "The sheer volume and variety of state climate initiatives is staggering" and appears to have been accelerating since
2008, owing to the continued inaction by the federal government (Rabe
2009, 72).

Initially, scholars were skeptical that these local, regional, and state
efforts would have much impact on reducing greenhouse gas emissions.
However, these efforts have been so abundant that they may add up to
significant changes. According to Lutsey and Sperling (2008), if cities' and
states' climate change goals were to be achieved, they would be equivalent
to a 47 percent reduction in the total U.S. emission reduction that would
have been required under the Kyoto Protocol had it been ratified by the
United States. The combined impact of these initiatives would actually
stabilize U.S. GHG emissions at their 2010 levels until the year 2020.

Even within the business sector, firms have chosen to reconsider an
initial anti-climate change regime position. Some businesses in the United
States have responded instead to the global movement to combat climate
change. As one scholar remarks, "If anything, corporate climate strategies became more diverse even in the US, and the ground started to shift
in favor of US engagement with international climate action long before
the end of the Bush administration" (Falkner 2010, 111). Some U.S. businesses have voluntarily adopted climate change controls. For instance,
the Chicago Climate Exchange (CCX) instituted a voluntary emissions
reduction and trading system that includes major corporations such as
American Electrical Power, DuPont, and Ford Motor Company. In addition, cities such as Chicago have pledged to reduce greenhouse gas emissions to 80 percent of 1990 levels by 2050 (Fisher et al. 2013). Other U.S.
businesses have begun to recalculate their long-term investment plans to

include climate change initiatives, even though U.S. political institutions have not yet mandated reductions.

These pressures from subnational states and municipalities have affected even those U.S. corporations staunchly opposed to climate change proposals. The United States, with its varying governments at the state, county, and municipal level, poses a potential headache for firms operating in a national market, encouraging corporations with a nation-wide market to push for uniform federal regulations. Similarly, firms with international markets have an interest in standardized international regulations that level the playing field rather than the creation of implicit non-tariff barriers (Porter and van der Linde 2000[1995]; A. Hoffman 2001). The current checkerboard of municipal, state, and county variations in climate change regulations promises exasperating complexities for corporate calculations. As Jones and Levy (2009, 235) note, "This uncertainty presents an obstacle for corporate planning."

Consequently, even in the United States, the successful negotiation of the Kyoto Protocol in 1997 "shifted expectations regarding future carbon restrictions and made climate-related business risks more tangible," despite the lack of ratification by the U.S. Congress (Falkner 2008, 111). Corporations that adamantly opposed ratification, such as those in the energy sector, have begun to demand a stable regulatory environment for climate change issues (A. Hoffman 2007, 2). In addition, some U.S. corporations began to fear that "the political gridlock over climate policy at [the] federal level might soon be broken. ... The changes in the political landscape have created a powerful rationale for hitherto reluctant business actors to embrace the idea of more progressive climate policies, as has become evident in recent corporate testimonies on Capitol Hill" (Falkner 2008, 135). Thus despite the obdurate resistance at the federal level, significant activity is occurring below the surface.

Clearly the problem of global climate change is far from being solved – and there is no guarantee that a satisfactory resolution will be attained. However, climate change issues have been institutionalized to the extent that they are unavoidably on the international agenda. Climate change is on the national agenda in most industrialized countries and is increasingly becoming a political issue in many developing countries as well (Giddens 2009). As Falkner (2008, 139) notes, "Overall, business has been unable to control the emergence of the global climate agenda" even though it "has significantly slowed down the rate at which international regulations have been adopted." As a consequence, the creation of the structure afforded by the Kyoto Protocol has enabled activity around the meaning and growing cultural salience of climate change.

When conflicts between institutions are in evidence, as in the case of climate change, there is no guarantee over which set of cultural meanings will prevail. Perhaps changing economic conditions will alter the institutional context, tipping the balance toward the adoption of one set of cultural meanings, or conclusive scientific evidence will create consensus around one position in the debate. The point of this section is not to claim that pro-environmental meanings will necessarily triumph. Instead, it illustrates that the growth of the cultural meaningfulness and salience of climate change has affected activity within the institutional field. The meaning of climate change has shifted considerably in both world society and within the United States over the past two decades. Although it is currently undetermined which set of environmental meanings will predominate in world culture, the world society perspective suggests that the growing meaningfulness of climate change discourses is a critical factor in the eventual effectiveness of the global environmental regime.

Conclusion

Drawing on a "thin" conception of environmental norms, the interest-based scholarly literature argues that business corporations are intrinsically motivated by their own interests and are unlikely to give up profits merely for the sake of pro-environmental norms. In contrast, the world society perspective suggests that cultural meanings embodied in institutional structures fundamentally reorient the activity of a society. Institutionalization does not imply the absence of political conflict or controversy. Instead, the macro-level world society perspective suggests that conflict ensues as an institution expands into new waters, altering the cultural meaning of the institution itself. At times, this expansion may lead to the agglomeration of cultural meanings with little overt fuss. At other times, however, this expansion can lead to conflict over institutional boundaries. This conflict typically signals change in both institutions rather than a clear victory of one over the other – although the final outcome is contingent and difficult to predict.

6

The Limits of International Institutions

This book has set out to understand the conditions under which international institutions generate large-scale social change. To fully address this question, one must consider the converse: when is social change unlikely to occur? There are many international efforts and social movements that aspire to change the world. However, in many cases, efforts to bring about social change have not borne fruit. Issues fail to get on the agenda, fail to generate policy reform, and ultimately have had little consequence for social practices around the globe. This is certainly true for many environmental issues, despite the overall growth of the global environmental regime. Examples include international efforts to address overconsumption, desertification, and deforestation.

This chapter first considers traditional arguments about the role that pro-environmental values play in motivating social change. Environmental activists often believe that attitudes and values – the virtuousness of their objective – are the key factor in bringing about widespread change. Frustration occurs when strenuous pleas for a good cause fall on deaf ears and apathetic bodies. In contrast, the world society perspective posits that cultural meanings are derived from existing institutions. Moral pleas themselves are made meaningful by a culture that is rooted in institutional structures. Consequently, institutional structures provide the primary impetus for the perception of virtue as a motivator of social change.

Second, the book's core arguments are mobilized to explain negative cases – instances where social change does not occur. Following the world society model, efforts to motivate social change are likely to fail in the following circumstances: (1) international institutional structures are not established, (2) institutions fail to generate the proliferation of agents, and (3) institutions do not transform cultural meanings. Mass consumption, desertification, and deforestation are briefly discussed as environmental issues in which little progress has been made through international regimes.

Third, the chapter outlines a world society explanation for revolutionary change. Incremental change provides the backbone for the arguments in the book. Nevertheless, revolutionary institutional change can occur when there is a significant restructuring of the international system. These restructurings may be the result of major wars, such as the Napoleonic Wars of the nineteenth century or the world wars of the twentieth century. The aftereffects of the Cold War have also led to some revolutionary changes as well. These wars can lead to the wholesale transformation of the international system, which in turn can produce dramatic social change. Finally, the chapter speculates on the tendency toward optimism among world society scholars.

Value-Based Motivations for Social Change

Efforts to bring about social change are based on an idealized vision of the way the world ought to be instead of the world as it is. Yet there is a critical distinction between world society theory and alternative perspectives on the role that attitudes and values play in motivating social change. In modernization and social movement arguments, values provide an underlying impetus for change. Objective knowledge about environmental degradation changes public attitudes and spurs efforts to improve conditions, either in society generally or via particular social movements. In the world society perspective, by contrast, institutional structures provide the basis for new cultural meanings with implications for action. Meaning itself is constructed through institutions rather than being self-evident to all or simply recorded by science. Consequently, value-based appeals that draw on existing structures will be more likely to motivate social change than will value claims unmoored from an institutional basis.

The dominant view assumes that values are the primary starting point for social change. From this perspective, the strength of the normative appeal is the critical factor that motivates social change. This *a priori* view of value-based appeals agrees with modernization and social movement perspectives. The modernization perspective claims that social institutions will tend toward increasingly efficient solutions. Thus the stronger the benefits of a proposed social change for society in general, the more likely that change will occur. For the sake of explication, this hypothesis is presented in a highly simplified form, omitting significant complications such as disagreements over what constitutes an improvement, definitions of which segment of the population is prioritized, and difficulties of implementation. However, the boiled-down essence of the argument suggests an analogue to Archimedes' lever: given a value-based appeal that is sufficiently strong, one person could move the world.

The social movement perspective similarly draws on the strength of public attitudes and values as the key factor in predicting social change. Akin to the modernization argument, social movement activism is rooted in the societal value of the proposed social change. However, some forms of social change may face additional obstacles. In some cases, the social value might chiefly benefit under-served populations that lack powerful advocates. Or a proposed change might benefit society broadly but work to the detriment of a few powerful opposing interests. In these cases, the normative claim must be amplified by the work of social movement groups in order to gain sufficient strength to attract political attention.

In contrast, the world society perspective suggests a fundamentally different role of value-based claims in the promotion of social change. Cultural meanings are important to the process of social change, as discussed in Chapter 5. However, cultural meanings do not provide the initial starting point, but must be embedded within institutional structures in order to bring about social change. This does not imply that new cultural meanings are never created – the development of changed environmental meanings since the 1970s clearly shows the opposite. However the social acceptance of new cultural meanings is more likely when meanings derive from existing institutional structures. Thus value-based claims are more likely to be the consequence of institutional structures rather than the initial motivation for their formation. Institutional change is not motivated primarily by the degree to which society would improve from a value-based standpoint, as the modernization perspective would expect. Instead, institutional change usually occurs by building on or extending preexisting structures and cultural meanings.

This world society perspective on the role of value-based claims fits with the broader social construction perspective (Berger and Luckmann 1980[1967]). From the social construction perspective, cultural meanings are allocated by social institutions rather than depending on eternal and universal truths. This suggests a form of cultural relativism, in which the values of a society cannot be judged against some absolute standard. Nor can we be sure that we live in the best of all possible worlds. Instead, normative claims only make sense within a particular set of social institutions. Consequently, value-based claims that derive from existing structures are likely to be more successful in promulgating social change than are claims that are not supported by institutions.

The Converse of Institutional Change

The theory of change developed in this book should also be able to explain why change fails to occur. According to this argument, social change is unlikely to occur in the absence of institutional structures, or without enabling agents or cultural meanings. These points are briefly illustrated with examples of overconsumption, desertification, and deforestation.

The Bee Swarm model poses a challenge for evaluating the extent of social change. Since the Bee Swarm model posits that change may occur in a roundabout fashion, the apparent lack of change in the short term does not necessarily imply that social change will not occur eventually. From this perspective, the impact of an international institution can only be evaluated over a substantial period of time, often measured in decades. This poses obvious challenges for empirical research. The cases discussed here involve little apparent progress, despite a couple of decades of international efforts – but one cannot rule out the possibility that change will occur in the future. Ultimately, the jury is still out. Nevertheless, this section will plunge ahead and analyze these as cases where global institutions have so far been ineffective in improving environmental conditions.

Structure: According to Chapter 3, the emergence of international structures for dealing with a given environmental issue increases the likelihood of social change, regardless of an absence of agents or cultural meanings at the outset. Conversely, the absence of structures at the international level implies that worldwide social change is much less likely. One example is the issue of overconsumption. Throughout the course of the modern environmental movement, mass consumption has been identified as one of the major contributors to environmental degradation. In the 1970s, concerns over overpopulation and resource depletion were based on the excessively high consumption levels of industrialized Western states (Ehrlich 1968; Commoner 1971, Catton 1982). In the 1980s, Schnaiberg and other scholars pointed to the dangers of the "treadmill of production" that required the overproduction of consumer goods in order to maintain production (Schnaiberg and Gould 1994; Gould et al. 1996). In the 1990s, the concept of the "ecological footprint" was developed to measure the amount of earth's resources taken up by each person (Rosa et al. 2010; York et al. 2003). Overconsumption has been a central theme in the cultural understanding of the environmental crisis in industrialized states.

While overconsumption has concerned environmentalists, it has not yet attracted much attention from major international organizations or national governments. One might hypothesize that corporate interests have blocked attempts to address the issue, but there is little evidence either of attempts to create structure or of corporate opposition to such attempts. Nevertheless, without international structures to address overconsumption, it has been difficult to generate widespread social change. As argued in Chapter 3, institutional structures play a critical role in defining and constructing social problems and placing them on national and global agendas. Thus far, there have been no intergovernmental conferences or treaties addressing issues of overconsumption. Nor has the issue made it onto national political agendas or been the focus of proposed national legislation. The basic work of creating standards and consensus on the problem of overconsumption has not taken place. In the absence of a global regime, there is little agreement on the basic contours of the problem, much less a conventional standard of acceptable consumption or consensus over what would constitute overconsumption.

Without an international regime and clear definition of the problem, coordinated action is unlikely. As Jepperson (1991) points out, disorder and heterogeneity rule in the absence of institutions. People adopt their own views on what constitutes environmentally sustainable consumption. Some people may choose to live "off the grid" and radically reduce their ecological footprint. Suburban families with minivans may also claim to have lowered their consumption by reusing grocery bags or using cloth diapers instead of disposable ones. Without institutional structures to embody and stabilize cultural meanings about overconsumption, it is difficult to even define the problem much less create pressure to resolve it. While "voluntary simplicity" or minimalist movements might be identified in some domestic cultures, these efforts tend to be labels for individual choices rather than coordinated social movements. Reducing consumption might be highly desirable in efforts to improve environmental protection. However, the lack of international and national structures focused on this outcome decrease the likelihood of substantive social change.

Agents: Social change is less likely to occur when structures exist yet there are few agents empowered or mobilized to work on the issue. Agents – in bureaucracies, social movements, and firms, facilitate the implementation of international and national regulatory structures, and otherwise prod states and firms to alter behaviors. It is not clear from a theoretical standpoint why some international regimes effectively mobilize hordes of agents while others fail to do so, although ad hoc explanations abound. However, the existence of structures is a precondition that greatly increases the likelihood that agents will be recruited eventually to work on the issue.

An example where a lack of agents has impeded social change can be found in international efforts at halting desertification – the expansion of desert due to the degradation of existing drylands. Much scientific evidence suggests that the economic and environmental benefits of slowing desertification would be substantial. Halting desertification would increase agricultural farmland and grazing area and potentially address global warming. The causes of desertification are fairly well known, including the over-cultivation of farmland, overgrazing of cattle, deforestation, and improper irrigation practices (Guruswamy 2007; Grainger 1990; Baker 1980). Easiest and cheapest would be the implementation of preventive measures for lands that have not yet become significantly degraded (Tolba and El-Kholy 1992, 147). Such measures might entail the reversal of deforestation, improved management of agriculture and grazing practices, and the use of improved technologies such as wood-burning stoves (Sitarz 1993, 103). While these changes would require financing from the international community, they are straightforward compared to solutions for environmental issues such as climate change that are technically complex and require substantial social dislocation. International efforts seem likely to bear fruit, since inexpensive technologies currently exist to halt desertification.

Moreover, significant international structure has been built around the issue. International efforts to control desertification began with the 1977 United Nations Conference on Desertification held in Kenya and the adoption of a UN Plan of Action to Combat Deforestation. Additional efforts at building structure were undertaken at the UN Conference on Environment and Development in 1992, resulting in the 1994 UN Convention to Combat Desertification. Desertification was also on the agenda at the 2000 UN Millennium Summit and the 2002 World Summit on Sustainable Development. A few social movement organizations, such as EarthAction, have sought to promote an anti-desertification agenda in the international arena (J. Smith 1997).

However, almost forty years later, relatively little substantive improvement has occurred in desertification outcomes. In contrast to the issue of deforestation, in which forests have been portrayed as a global commons, the expanding deserts of large stretches of land in Africa, Latin America, Asia, and North America, have garnered little public attention. There is no obvious value-based reason why desertification has not generated more action. Nor are oppositional institutions evident as in the case of climate change and deforestation. Instead, there has simply been a relative dearth of agents taking up the issue. The usual array of international agents has not emerged, such as INGOs, scientists, government leaders in developing countries, and activists. Nor is there much mobilization around desertification among farmers, agronomists, foresters, and planners of countries with expanding desertification (Grainger 1990). Interestingly, some of the more recent efforts of international organizations are aimed at the recruitment of such agents. For instance, one scholar describes the 1994 UN Convention to Combat Desertification as adopting "an innovative 'bottom-up' approach" to desertification as a way of enlisting agents from developing countries for this task (Guruswamy 2007, 567).

The case of desertification suggests that institutional structures alone are insufficient to bring about social change. Agents must also be empowered to promote social change, either by institutional structures or perhaps by social movements outside of the structure. The emergence of preliminary international efforts to address desertification is promising. For instance, desertification is on the UNEP agenda, and there is now substantial consensus about the definition of the problem, as well as various proposed solutions. The outlook for social change is, consequently, more promising than in the case of overconsumption. Yet the case of desertification highlights the fact that international structures, alone, are insufficient to generate change.

Cultural Meanings: Even the emergence of international structures and agents may be insufficient to generate social change if commonly agreed-upon cultural meanings have not been established. According to the Bee Swarm model, change occurs when myriad forces push in a consistent direction. If new cultural meanings fail to develop, various forces will push in heterogeneous directions. When convergence in cultural understandings does not occur, social change is less likely.

Failure to establish consistent cultural meanings is exemplified in international efforts to decrease deforestation, especially in tropical countries. This issue is near the top of the global environmental agenda, as deforestation is closely implicated with other global concerns such as climate change and biodiversity (Cadman 2011; K. Brown and Pearce 1994; Hecht and Cockburn 1989; Humphreys 2006; Vajpeyi 2001; Bunker

1985; Rudel 2005; Rudel 1993). Global deforestation has been the subject of several international treaties and conferences, including the 1992 UN Conference on Environment and Development, the Intergovernmental Panel on Forests and its successors, and the 1995 World Commission on Forests and Sustainable Development. Because of the prominence of the issue on the international agenda, a large number of environmental agents are working on behalf of deforestation within the international community, in Western states, and in developing countries.

However, the issue of global deforestation is currently a case in which conflict and opposing meanings have prevented the emergence of common cultural understandings regarding forests. On one hand, within the ecosystemic environmental perspective that is central to the post-Stockholm regime, forests are understood as the "lungs of the earth," which can slow the effects of climate change as well as provide habitat for irreplaceable species. From this cultural standpoint, rainforests are part of the "'global commons' and represent a collective good" that it is in the interest of all states to protect (Hurrell 1992, 401). This view of forests is a relatively recent cultural meaning, emerging primarily in the 1980s. While it has been largely adopted within the international arena and in most (but not all) Western states, this perspective remains contested in many of the developing countries with large forested areas.

Opposition to the ecosystemic perspective on forests derives from an alternative institutional structure rooted in the earlier resource management perspective. From the resource management perspective, forests are viewed primarily as an economic resource for the nation-state rather than as a global good. Yet it would be a mistake to assume these meanings follow self-evidently from economic interests. Instead, considerable effort was undertaken in the early twentieth century to develop institutional meanings and structures that encouraged sustainable management practices to maintain a supply of harvestable trees (Hayes 1959; Nash 1973). The institutionalization of the resource management perspective at the world level can be seen in treaties such as the1984 International Tropical Timber Agreement and organizations such as the World Bank (Humphreys 2006). Interestingly, some recent efforts at the international level have focused on the development of forest certification schemes that focus on providing more environmentally sensitive standards for forest management (Cadman 2011; Humphreys 2006). These schemes tap into resource management cultural meanings while also creating potential synergy with the ecosystemic values of forests as a planetary resource.

Despite the considerable amount of international activity, however, the opposition of meanings embodied in different institutional structures has prevented broad consensus within the international community. Consequently, progress on deforestation can only be considered haphazard. While forest area has expanded in several Western countries, the enormous forests of tropical countries such as Brazil, Indonesia, and elsewhere continue to diminish (Myers 1992). Nevertheless, the existence of institutional structures as well as agents working on the issue encourages a prediction that change is likely to occur on this environmental issue, albeit in a Bee Swarm fashion.

Revolutionary versus Incremental Change

The world society arguments developed in this book have focused on incremental forms of social change. Not surprisingly, incremental change tends to be the more common type of change that occurs. However, world society theory can also account for revolutionary forms of social change. Revolutionary change from the world society perspective follows from a radical restructuring of the international community. The radical political transformations that accompany a major world event are the most likely precipitators of radical restructuring of the international system. In contrast, modernization and Marxist theories posit economic forces as the primary drivers of revolutionary change.

Revolutionary change is comprehensive, allowing previously unthinkable social change to occur once existing institutional structures are swept away. In contrast, incremental change takes place within the parameters of existing social institutions. Following the arguments in this book, incremental changes tend to occur when new institutions are formed based on preexisting structures. However, as argued previously, these incremental changes are not predetermined. Agents, cultural meanings, and contingent events influence the speed, style, and content of the expansion of institutional structures. Social change occurs as institutions implement, expand, and innovate. Although the modern global environmental regime is a new institution, it derived from preexisting structures in the international system, as argued in Chapter 2.

Yet radical changes in social structure also occur. These occurrences are akin to seismic shifts in the international system that fundamentally affects institutional structures. In the scholarly literature, most revolutionary changes have been theorized as the result of major economic shifts. Both modernization and Marxist theories posit that broad economic transformations from agrarian to industrial or postindustrial societies provide the foundation for major changes in society, although the theories differ in the mechanisms by which these revolutionary changes occur. A corollary posits that major changes in technology, such as revolutions in transportation or communications, may also provide the instigation for major social changes.

Revolutionary political changes might also lead to major institutional change as well as economic ones. Such changes might follow the end of a major world war or other cataclysmic shift that substantially reconfigures the international system. Such events lead to new international structures such as the United Nations, new states emerging as Great Powers, and concomitant changes in the cultural meanings of world society. These events are external to the dynamics within environmental regimes, yet they can significantly shape the formation of environmental institutions and affect environmental outcomes.

For instance, world society scholars have explained institutional structures as an expansion of the ideas of the Enlightenment (J. Meyer 1980; J. Meyer et al. 1997a). In the late eighteenth century, French philosophers developed discourses on human rights that were quite revolutionary in the age of the absolute monarchs (Sewell 1980). The ideas themselves have had lasting effects on the expansion of the state, the legitimation of the individual, and concepts of democracy and justice. Yet these ideas were supported by major structural shifts that accompanied the French Revolution and the ensuing Napoleonic Wars. The international system at the conclusion of the Napoleonic Wars in 1815 was far different than it had been in 1789. New states had been made while others had been dismantled, France was the new world power, and democratic liberalization began to occur in all of the European states (Hobsbawm 1964). Within this radically restructured international system, substantively new institutions could be developed.

Other revolutionary shifts accompanied the ending of the First and Second World Wars. The end of the First World War led to the creation of the League of Nations, which presided over a similarly altered international system. However, this system was short-lived, as the onset of the Second World War led to the creation of yet another international structure. World society scholars have often dated the formation of particular institutions to the end of the Second World War (J. Meyer et al. 1997a; J. Meyer et al. 1997b; J. Meyer 1980). These new institutions were made possible by the formation of the United Nations system as well as Anglo-American biases in the international structure that emphasized liberal values, civic participation, and other Western cultural values. In an alternate world, the victory of Nazi Germany or the worldwide dominance of the Soviet Union would have led to the institutionalization of a considerably different set of cultural meanings within the international system. The restructuring of the international system following the Second World War both enabled and constrained the formation of the international environmental structure to develop in particular directions.

Another radical shift has been the end of communism in Eastern Europe. Although the Cold War was not technically a war, its aftermath led to the restructuring of the international system much like that of a world war. New states were formed, old states were dismantled, and the ideology of Marxism has been largely delegitimated in world society. The magnitude of this change has been less perceptible in the Western world, because Western cultural values have come to predominate throughout the whole world. Nevertheless, the fall of communism led to major changes in the international system that, in turn, allowed institutional changes that were unlikely in the world of the Cold War.

These radical shifts in international structure allow new institutional structures to arise that did not derive from previous but now-defunct structures. However, although revolutionary shifts allow a system to be restructured, that does not imply that all proposed new institutions will be allowed to flourish. Instead, the new institutions will be constrained and enabled by the new international structures. Nor are these new institutions determined solely by the new international system – agents, cultural meanings, and contingency play essential roles as well. In short, revolutionary changes in the international system allow concomitant revolutionary changes in institutions, yet those institutions are still subject to the processes of social change outlined in this book.

Optimism and Pessimism

The world society perspective is generally perceived as adopting an optimistic perspective on the likelihood of social change, and the arguments in this book are likely to be viewed similarly. Perhaps this appearance of optimism stems from the tendency of world society scholars to focus on the institutions in the global community that promise to promote progress and justice. However, the world society perspective is fundamentally agnostic about whether progress is truly being achieved. Similarly, this book has sought to formulate a theoretical explanation for the types of change that are empirically evident in society. Yet the explanation of the changes that do occur in society should not be taken as implying that ideal progress is being made in the protection of the environment. This section examines two serious caveats to environmental optimism.

The first caveat is that environmental degradation is still occurring. The expansion of structures, agents, and cultural meanings promoting environmental protection at the global level are likely to have an ameliorating effect, at least on those issues that have led to the formation of institutional structures. In a few cases, global attention has led to near-complete reduction in harmful environmental practices, such as the wholesale dumping of pollutants in the ocean and the production of chlorofluorocarbons. However, in most cases, global institutional processes have at best reduced the pace of environmental degradation.

Consequently, existing efforts of environmental protection may prove insufficient to address the magnitude and urgency of environmental problems. Pollution and resource scarcity are growing apace with economic development and population growth. Even if these environmentally damaging processes are slowed by the global environmental regime, degradation is still occurring. Even more alarming, certain natural processes may reach a tipping point. Scientists have theorized that once the earth's temperature has risen to a particular level, reinforcing dynamics will render climate change irreversible (National Research Council 2010; Intergovernmental Panel on Climate Change 2007). For these processes, the slowing of environmental degradation may be insufficient to prevent ecological catastrophe.

The second caveat is that international institutions may be targeting the wrong problems or directing the wrong sorts of changes. Change is likely to occur along the directions indicated by global structures, albeit in a slow and roundabout fashion. However, these institutions may not be the optimal institutions for environmental protection. Instead, there is the real possibility that global institutions have identified the wrong issues or have set the wrong standards for amelioration. Perhaps a looming environmental disaster even more catastrophic and imminent than global warming exists, but has not been taken up by the global environmental regime.

Such possibilities are difficult to plan for since they are, by definition, unforeseen. However, this tendency to bark up the wrong tree is evident in economic and political arenas, where decision makers often focus on one threat while failing to notice others that ultimately bring about disaster. In the environmental arena, this may be illustrated by the failure to build institutions that address overconsumption associated with capitalism, which is a primary source of environmental degradation. The same pattern can be observed in military affairs. The theories of military strategy of one era are quickly replaced after unforeseen military debacles.

Ultimately, the fate of the environment is unclear. The global environmental regime has generated a variety of improvements in environmental outcomes. However, in only a few cases have environmentally destructive practices ceased. In most cases, the production of environmentally damaging products continues at a somewhat dampened rate. Although institutional structures are in place, environmental agents need to carry out a great deal more work in order to produce globally sustainable environmental protection.

7

Conclusion

The first wave of world society research in the 1980s and 1990s focused on global diffusion and policy isomorphism. Scholars observed that international institutions could have dramatic effects on government policies in areas such as education, human rights, and the environment. However, these early studies left open the question of whether global institutions would lead to actual improvements in human rights, educational practices, or environmental protection. Policy isomorphism seemed likely to generate a large gap between policies and social practices, particularly in poor countries that lacked the resources for effective policy implementation. The observed disconnections between policy and practice had actually inspired the formulation of world society theory in the first place. Education policies in Sub-Saharan Africa, for instance, appeared to reflect global ideals rather than local needs, departing from standard functionalist views. These observations motivated early claims that state behavior reflected the scripts or myths in global culture rather than functional efforts to address local problems.

Early world society scholarship intentionally focused on empirical instances of loose coupling between policy and practice as a way to refute arguments that state policies and structures could be explained as rational behavior or functional adaptation. Decoupling and obviously dysfunctional policies created the space for the claim that global culture mattered. While this strategy helped establish world society as a viable alternative to modernization theory and various rationalist perspectives, it had some unfortunate consequences. Subsequent empirical research was channeled toward the analysis of discourse and policy, with relatively less emphasis on social practices and social change. Scholars were more apt to focus on symbolic isomorphism – for instance, the adoption of national flags and anthems – rather than investigating the real-world consequences of rapidly diffusing health, education, welfare, and environmental policies. This contributed to a common misconception that world society theory was a theory of "myth and ceremony," rather than real behavior (Schofer et al. 2012).

This book serves as a theoretical and empirical corrective, exploring how international institutions transform social activity around the globe – yielding tangible consequences for the natural environment. Recent scholarship offers an apparent paradox: statistical evidence finds that international institutions generate real-world social change, but case studies frequently criticize particular international treaties and regulatory structures as loosely coupled and ineffective. Chapter 1 squares this circle by articulating a "Bee Swarm" model of social change. International institutions may not always have direct effects on substantive outcomes, but they nevertheless prompt social change via a multiplicity of mechanisms. Institutional structures set agendas, establish workspaces, and create incentives to work out solutions. Institutional structures also legitimate and empower agents to address these issues. And institutions embody and propagate new cultural meanings that support action. Consequently, institutions in world society can generate social change even if their formal regulatory structures and sanctions are weak.

This is not to say that international institutions are wholly effective. As the empirical cases and examples in this book demonstrate, a host of organizational, cultural, and political dynamics influence the formation and impact of workspaces, agents, and cultural meanings. Comparative analysis of different institutional sectors would advance the literature, both by interrogating these dynamics further and by determining how and when they prompt widespread social change.

Origins

Chapter 2 examines the emergence of the global environmental regime. Conventional accounts of the 1970s often stress the role of popular movements or heroic individuals such as Rachel Carson (but see D. Meyer and Rohlinger 2012), while case studies of international efforts tend to focus narrowly on the creation of particular international treaties or organizations (Benedick 1998; Brown Weiss 1998; Pring 2001). This book develops an alternative story, in which older institutional structures and cultural models were reassembled into a novel package that proved highly generative. In doing so, the chapter addresses a long-standing criticism that world society theorists have paid insufficient attention to the origins and emergence of institutions.

Why did the global environmental regime emerge? The challenge of tracing origins is that successful institutions appear inevitable in hindsight. The very success of the global regime served to establish new cultural meanings and logics for environmental action. Once that has been accomplished, it seems obvious that various social actors would perceive environmental problems and mobilize to address them. In other words, contemporary cultural understandings are easily projected into the past, effectively obscuring the origins of an institution.

The second challenge of studying institutional emergence is the dominance of actor-centric accounts. If one begins with the assumption that individuals and social movements of the 1960s and 1970s held some approximation of contemporary views on environmental problems, questions of cultural meaning and social construction are pushed to the background, while political and organizational issues are brought to the front. This focus on actors shifts explanation toward the environmentalists who fomented organizational or political change and the interests that opposed them. In fact, neither environmentalism nor the opposing interests were well consolidated before the formation of global institutional structures.

To avoid this potential for teleology, Chapter 2 reframes the question: How did the particular package of concerns, now labeled environmentalism, get assembled and become established at the international level? The counterfactual possibilities include the failure to become institutionalized, but also alternative ways of conceptualizing and packaging environmental concerns. The modern environmental regime reflected innovations that were as much cultural as organizational. The Stockholm Conference and its aftermath served to consolidate the contemporary understanding of environmental problems, linking formerly disparate issues under a single umbrella. Prior international environmental efforts had much more limited success, and one can imagine counterfactual institutional configurations that would have also addressed the different combinations of issues related to the natural world.

The account of institutional origins in Chapter 2 is both structural and contingent. Prior institutional structures, including early environmental treaties and the UN system itself, provided both opportunities and constraints that influenced the environmental regime that developed at the Stockholm Conference. However, the global environmental regime was not simply the logical or inevitable extension of prior structures. The

Stockholm Conference represents an origin point because it involved cultural innovation and the emergence of new international structures that proved highly generative of subsequent activity.

The story begins with institutional precursors to the Stockholm environmental regime, including preservationist and resource management institutions that developed in the nineteenth century. Like the contemporary regime, these earlier institutions included cultural and organizational dimensions. Both traditions entailed distinctive cultural understandings of nature that became embedded in several international organizations and treaties, and empowered a variety of agents including scientific experts and early social movement groups. These served as cultural and organizational building blocks for the post-Stockholm regime, and agents linked to early conservation movements were key players.

The nascent environmental regime could have evolved in many ways. One likely outcome would have institutionalized the resource management framework as the dominant understanding for environmental problems. Another outcome might have parceled various issues among existing international institutions: for instance, the OECD might have taken up the pollution concerns of the industrialized nations, WHO might have focused on the impact of environment on health, and other international organizations might have divided up other issues. The emergence of UNEP, an independent agency of the United Nations, was by no means necessarily determined.

Chapter 2 explains the emergence of UNEP as a contingent outcome. A variety of proto-environmental conferences occurred, or were planned, in the years surrounding Stockholm. In hopes of engendering greater attention and participation, the Stockholm organizers – working as entrepreneurial bureaucrats rather than social movement activists – sought to widen the range of topics addressed in the conference. In the process, they forged a broader cultural frame for thinking about environmental problems that linked a previously diverse set of issues. This new frame, which evolved to some extent in the years building up to Stockholm and afterward, effectively brought countries of the developing world to the table and made possible the creation of UNEP. The organizational structures of UNEP, in turn, helped to institutionalize the broad Stockholm vision of environmental problems. The resources and legitimation of the new UN agency, in combination with a potent new cultural frame, ultimately proved to be highly generative. In the following years, international pro-environmental activity accelerated rapidly around this new regime.

Chapter 2 has several implications for future research. Studies that focus on origins may benefit from the macro-level perspective employed by world society scholars. Focusing narrowly on any single event in the process of institutional emergence is likely to result in ad hoc explanations emphasizing key actors and their strategic maneuvering. In contrast, Chapter 2 stresses the need to examine a broad range of possibilities for institutional formation – the multiplicity of cultural logics, international conferences and events, and early governance regimes – rather than only those that ultimately succeeded. Given this set of opportunities, the founding of a particular global institution must be seen as probabilistic and contingent rather than inevitable. The challenge is to understand the historical confluence of events that led to the construction of the specific set of institutional structures, agents, and cultural meanings, in contrast to plausible alternatives.

The word origin is itself problematic, as new institutional forms are typically built on the foundations of prior ones. Efforts to identify the "first" origin result in a recursive search into the distant past. In this light, the account of the origins of the global environmental regime in Chapter 2 is better characterized as one significant moment in a continuing series of formative events. Imagining institutional structures as malleable and continuously re-formed obviates the need for dramatic accounts of institutional origins or heroic actors.

This conception of origins leaves room for a degree of contingency. The world society account developed in Chapter 2 is both messier and more mundane than many accounts of institutional origins. Agents often build new institutional structures through the employment of ordinary bureaucratic routines rather than through revolutionary defiance. However, these agents are not mere automatons following marching orders from a monolithic global culture. Bureaucrats, scientists, and various organizational participants are important environmental agents, employing considerable skill in the implementation, translation, and innovation of institutional routines. These efforts are most often effective when employed to extend or reformulate preexisting institutional components, rather than seeking to radically depart from existing structures.

In sum: the real origin story is in the assembly of a new package – a combination of cultural and organizational forms that develop into full-fledged institutional structures. World society accounts of institutional origins should seek to identify the prior institutional structures and meanings, and the contingent processes through which institutional agents generate and consolidate new institutional forms.

Institutional Structures

Chapter 3 tackles the question of how international structures generate social change. Case studies of particular international environmental treaties or organizations have frequently revealed lax environmental standards, poor national implementation, and toothless enforcement. Not surprisingly, there is a fair bit of pessimism in the literature regarding the possibility for effective global environmental governance. The assumption that institutional structures operate via direct effects – the Smoking Gun model – means that poor implementation and enforcement will doom efforts to solve environmental problems. Chapter 3 looks at an alternative, the Bee Swarm model of social change, which explores the multiplicity of indirect influences that are generated by institutional structures. This perspective explains how substantive improvements in environmental conditions can be observed despite the conspicuous flaws of international pro-environmental initiatives.

How do institutional structures matter? Rather than focusing on the direct effects of international treaties and policies, Chapter 3 examines how institutions create a workspace where social problems are constructed, key actors are brought together, and potential solutions can be hammered out. The chapter also unpacks mechanisms through which institutional workspaces propel social change. For instance, institutional workspaces facilitate the assembly of new knowledge and expertise – which serves as the foundation for constructing new environmental problems. Moreover, workspaces empower new environmental agents that may work toward new solutions. The resulting solutions are rarely optimal and sometimes terribly inadequate. Nevertheless, institutional workspaces facilitate the accretion of baby steps that over time may result in substantial improvements in environmental conditions.

Chapter 3 also identifies directions for future research. First, additional studies of the consequences of institutional structures are needed. The book's argument – that international institutions can matter even if policy implementation is weak – has been demonstrated in a few sectors of international governance. The extent to which this argument can be generalized is deserving of study. A macro and longitudinal approach is needed to examine the consequences of world society, since effects may not operate through direct policy mechanisms of any given treaty or policy initiative. To observe the effects of the Bee Swarm, effective research designs must look for a long-term relationship between the development of international structures and empirical consequences on the ground. One effort to do this for environmental issues is presented in the Appendix.

Second, given evidence that international institutions are consequential in a given sector of activity, further scholarship is needed to explore the dynamics and mechanisms. Chapter 3 shows that in the case of the environmental regime, one critical dynamic is the establishment of a workspace within which social problems are constructed, agents are empowered, and relevant parties are brought together. Future research could assess the generalizability of these processes across domains of international activity and identify complementary processes in greater detail.

Agents

The world society perspective characterizes key participants – such as INGOs, social movements, bureaucrats, and policy professionals – as *agents* empowered by institutional structures (J. Meyer and Jepperson 2000). This provides an alternative to the conventional imagery of interested actors that dominates the social sciences. World society theory draws attention to the ways in which institutional agents are empowered by broader structures. The emergence of the international environmental regime encouraged the formation of new environmental agents – UNEP bureaucrats, social movement groups organized under the post-Stockholm framework, new scientific research groups funded by international environmental institutions, and throngs of other types of agents. Meanwhile, preexisting environmental agents became much more effective in promoting environmental protection due to the support and legitimation of the expanding environmental regime.

The arguments in Chapter 3 offer a different lens for understanding the role of social movements in promoting social change. Rather than assuming that social movements are an exogenous force – an alternative explanation to world society arguments – Chapter 3 argues that social movements and other environmental agents are important components of institutional dynamics (also see Tsutsui and Shin 2008). The world society perspective goes further in stressing the embeddedness of agents within world society structures (see Longhofer and Schofer 2010). Additional research is needed to explore the relationship between emergent global regimes and the proliferation of agents and movements, as well as the impact of international institutional structures on social movement effectiveness. Chapter 3 argued broadly that institutional agents become more effective when empowered by international institutional structures rather than working outside of them. Even in the Western countries where social movements predated the growth of environmental legislation and structures, social movement effectiveness was magnified with the growth of pro-environmental institutions.

World society theory further departs from social movement scholarship by examining a wide variety of agents beyond social movements – such as environmental scientists, bureaucrats, and a host of other agents. Social movements may have influence in some political contexts, but pro-environmental change may also occur without the influence of any social movement groups. Further research might seek to identify the domains in which social movements are more important and, alternatively, the domains where INGOs, scientists, or other types of environmental agents have more traction.

In sum, the term agent is introduced as an alternative to the actor-centric imagery in most historical and social movement accounts of early environmental initiatives. The world society view builds on the common intuition that social change occurs more easily within supportive institutional structures. The idea that pro-environmental movements changed the political system from the outside is not consistent with historical evidence. Indeed, movements played less of a role in the rise of global environmentalism than is often imagined. Rather, agents – often bureaucrats and professionals – working within conventional arenas proved central to historical accounts and ultimately generated dramatic pro-environmental social change.

Cultural Meaning

Chapter 5 explores the role of culture in international institutions. World society theory draws on a relatively "thick" conceptualization of culture, as foundational meanings and understandings that empower agents and reshape interests. This contrasts scholarship focusing on "thin" cultural norms that come up against the *a priori*, obdurate interests of actors. The latter formulation sets up a false dichotomy between norms and interests.

World society stresses the role of institutions in stabilizing and propagating particular cultural understandings, including the post-Stockholm view of environmental issues. New institutional structures embody new cultural meanings. Moreover, institutionalized meanings are sometimes reshaped through the activities of agents and the influence of other institutions. For instance, the emergence and expansion of the human rights regime created new cultural meanings that may affect understandings of environmental issues.

New cultural meanings, in turn, shape the interests of actors. As an institution changes – with the expansion of structures, proliferation of agents, and changing cultural meanings – interests are transformed. Future world society research might map out the evolution of interests over time, as Andrew Hoffman (2001) does for corporate environmental interests over four decades. In addition, future studies should examine the interaction of multiple institutions and the consequences for cultural meanings and interests. Previous studies have described the changing meanings in single institutions such as the environment, women's rights, or human rights (Fourcade 2011; Chabbott 2003; A. Hoffman 2001; Berkovitch 2002; Barrett and Kurzman 2004; Barrett and Frank 1999; Heimer 1999; Kymlicka 1996). Yet these studies have tended to neglect the role that interactions with competing institutions have played in the evolution of cultural meanings. Future studies might explore the changing contours of an institution as it runs up against or incorporates aspects of other institutions.

World Society Theory: Considerations and Clarifications

This book has sought to clarify several of the intersections between world society theory and other theoretical perspectives in sociology. World society theory provides a novel perspective on society that diverges from commonsense intuitions. This allows new insights and encourages the observation of empirical phenomena that are disregarded by other theoretical perspectives. However, the drawback of novelty is that world society theory can be difficult to explain, allowing misunderstandings to arise. This section addresses questions about (1) heterogeneity, (2) the role of power, and (3) the importance of mechanisms from the world society perspective.

Heterogeneity

World society researchers have often painted with a broad brush, identifying isomorphic trends across a hundred countries at one stroke. As a result, world society research provides a contrast with historical and comparative scholarship that carefully unpacks the details of policy specifications, negotiations of interest groups, and ebbs and flows of public opinion within particular national or local contexts. Unfortunately, the empirical focus on isomorphism has led some to infer that world society theory cannot address heterogeneity of practices across peoples and places.

To the contrary, world society scholars are well aware that social life varies a great deal around the globe. Indeed the appreciation of comparative variability is what makes isomorphism, to the extent that it can be observed, so interesting. The concept of loose coupling helps make sense of persistent heterogeneity in a world that may nevertheless be increasingly isomorphic. According to the Bee Swarm model, social change results from myriad influences applied asymmetrically across the field. The unevenness of pro-environmental influences, in addition to variability in local context, results in heterogeneity across the organizational field. In many cases, these variations will themselves alter with time, perhaps leading to different configurations of heterogeneous variation.

Nevertheless, the world society perspective attends to the limits of heterogeneity. Global institutional structures stabilize cultural meanings of the environment, exerting a consistent pressure over time. While heterogeneous practices are likely to always be in evidence to some degree, they are limited by continual dialogue with the institutional structures and meanings at the global level (Strang 2010).

Power

Power is often invoked as an explanation for outcomes; yet when examined closely, the workings of power can be evanescent. Often, the concept of power functions primarily as a shorthand for the interest-based model of social change. Rather than assume interests are the primary causal factor, scholars might find greater analytic utility in examining the interests involved and the dynamic processes discussed in the section on interests in Chapter 5. The invocation of power often signals a post hoc label applied to known outcomes and provides no additional insight into the processes.

World society scholarship has generally focused on domains in which coercive forms of power are rarely invoked. This is the case for the global environmental regime, which has not been an arena for military conflict, and even diplomatic sanctions for environmental treaty violations are typically mild. However, even in arenas of coercive power such as interstate war, the concept of power is usually applied tautologically. Even in these cases, the victor of a war is labeled as powerful – power becomes a descriptive term rather than an explanatory one.

Instead, the application of power in world society scholarship tends to be more akin to the Foucauldian conception of power (Foucault 1990).

This power takes the form of setting the agenda, defining concepts, and providing the motivation for self-discipline. Moreover, institutional resources can be pivotal in influencing outcomes. Those interests that can afford to fund scientific studies, provide organizational resources and expertise, and engage large numbers of agents are likely to influence outcomes more than are interests with less extensive institutional resources. The institutional rules governing the allocation of these resources merit further scholarly study. However, grandly sweeping these factors under the label "power" obscures rather than enhances greater understanding.

Mechanisms

The tendency for world society scholars to undertake cross-national quantitative analyses has also tended to obscure the mechanisms by which world society influences are translated to national polities. In this case, the inattention paid to mechanisms by world society scholars belies a more profound disagreement on the theoretical importance of the identification of mechanisms in explaining social change. The close study of mechanisms is most reasonable in the Smoking Gun model, where policies are tightly linked to substantive outcomes. In these cases it may be productive to identify the specific mechanisms that effectively lead to social change, since such a mechanism may prove reliable for other issues and in different contexts. Frequently, however, advocates of the study of mechanisms are hoisted by their own petard. Although they can carefully trace the causal sequence of a particular mechanism, they are likely to be frustrated in attempts to identify its relevance to the outcome based on the assumptions of the Smoking Gun model. Since myriad other influences will simultaneously be affecting the outcome, scholars may be unable to conclude that the mechanism under scrutiny had a critical effect on the outcome.

In contrast, the tenets of the Bee Swarm model imply skepticism that the utility of a mechanism is generalizable, given the wide variation of structures and contexts across countries. Different countries are likely to have quite different political and economic contexts. Consequently, agents in different contexts are likely to utilize different mechanisms to attain the same end despite variations in their policy environments (Dobbin 1994; Boyle 2002). The Bee Swarm model leads to the expectation that a particular mechanism is likely to show only a weak causal influence in most cases. Weakness does not necessarily imply that the mechanism is unimportant, since the aggregated effect of all the weak influences eventually results in substantive change. Yet weakness does suggest that the utility of a mechanism is not necessarily generalizable to other conditions or issues.

Moreover, the emphasis on mechanisms overlooks the importance of agency. The world society perspective claims that environmental agents are constructed by institutional structures. However, even though their motivations and roles are socially constructed, these agents work creatively and strategically to carry out institutional goals (Fligstein 1997, 2001). Agents adapt environmental policies to different local conditions, utilizing different mechanisms depending on the institutional context. Agents in one location may be more skilled at adaptation than agents in structurally similar locations elsewhere. Mechanisms that work in one place may owe their effectiveness to the capabilities of the agents involved. Consequently, the same mechanism may have different effects depending on the skill of the relevant agents.

Since the world society model posits that a wide variety of mechanisms might be utilized in pursuit of environmental protection, with varying degrees of success, little attention is paid to country-specific mechanisms in this book. Instead, the tenets of the Bee Swarm model suggest that a variety of mechanisms are likely to be effective in motivating environmental change. Conversely, a mechanism that effectively promotes environmental protection in one economic and political context is likely to be ineffective elsewhere.

Conclusion

World society theory is often characterized as a theory of diffusion celebrating myth and ceremony. In fact, it embodies a general theory of social change. This book seeks to elaborate key parts of world society theory, including the processes through which new global institutions emerge, and the multiple ways that institutions prompt real changes in interests and activity on a global scale.

The chapters develop alternative concepts and imageries to the conventional actor-centric theories that dominate contemporary social science. Social change is the product of a Bee Swarm, rather than the direct consequence of a particular treaty or potent social movement. Institutions have consequences indirectly, by establishing workspaces, empowering agents, and transforming cultural meanings and interests, rather than simply opposing or counteracting the *a priori* interests of social actors.

Proponents of social change are often discouraged by the seeming obduracy of social problems evidenced by inefficient and corrupt organizations, powerful opposing interests, and disengaged publics. Nevertheless, a broader historical perspective demonstrates that societies change dramatically over time. Perhaps the incongruity can be resolved through the lens of world society theory, in which the meandering route of international institutions becomes the non-intuitive route to meaningful social change.

Appendix

The Tangible Effects of International Environmental Institutions

This book explores how international environmental institutions produce social change. Chapter 1 introduced the Bee Swarm model, in which social change is brought about through multiple indirect mechanisms. But this issue is important only if environmental institutions actually do have real-world consequences on the environment. In fact, a growing quantitative empirical literature has found evidence of these effects. Recent studies observe that international pro-environmental institutions and national ties to international environmental organizations are statistically associated with improved environmental conditions. For instance, studies have examined the long-term effects of institutional structures and linkages on pesticide use (Shorette 2012), recycling (Hadler and Haller 2011), water biochemical oxygen demand (Shandra et al. 2009, 2008; Jorgenson 2007), biodiversity (Shandra et al. 2009b), methane emissions (Jorgenson 2006), chlorofluorocarbon production (Schofer and Hironaka 2005), carbon dioxide emissions (Jorgenson 2007; Schofer and Hironaka 2005), and deforestation (Shandra et al. 2009a; Schofer and Hironaka 2005).

Given that the book hinges on this empirical point, it is useful to provide an overview of the statistical evidence showing the tangible consequences of international environmental institutions. This Appendix presents statistical analyses showing that international institutional structures and agents are associated with improved environmental conditions. This basic finding is extremely important, especially in light of the prior case study literature, which has largely focused on the failures of environmental governance.

Nevertheless, appropriate caveats are in order. The fact that international institutions "matter" is not the same as claiming that these environmental problems have been solved. Institutional structures and agents may have meaningfully dampened the rate of environmental damage being inflicted. However, dampening a problem is different from solving it. In only a few cases, such as the production of chlorofluorocarbons and the mass dumping of marine pollution, has there been a substantial absolute reduction in environmentally harmful practices.

Another note of caution is that the development of international pro-environmental institutions is uneven. Many important environmental ills have not yet been addressed by the global environmental regime (see Chapter 6). The institutional processes described in this book are only brought to bear when institutions have been constructed. To the extent that international environmental institutions remain weak or non-existent in many specific issue-areas (e.g., desertification), environmental problems may continue to accelerate.

With those caveats in mind, the core issue here is to present basic evidence that international institutions do matter – that they have a tempering effect on rates of environmental degradation. The research design is necessarily macroscopic. In a complex and loosely coupled world, policy initiatives may fail in particular locales or in the short term. To properly assess the effects of international institutions, one must look at the big picture: trends across many countries and over long periods of time. Does the emergence of international institutions and environmental agents correlate with improved environmental practices?

Data and Methods

This appendix uses cross-national longitudinal data to examine the consequences of international pro-environmental institutions. The empirical strategy is to search for a statistical association between international institutional structures and quantitative measures of environmental conditions such as pollution emissions or deforestation. This requires development of a statistical model to hold constant potential confounding factors that also influence environmental damage. These analyses are necessarily limited to sources of data that are available for large numbers of countries and are comparable over time.

The following analyses examine the effects of institutional processes on pollution and land use. Four types of pollution are examined: (1) carbon dioxide emissions, a major greenhouse gas that is a by-product of fossil fuel use, (2) sulfur dioxide emissions, a chief component in acid rain,[1] (3) nitrous oxide emissions, another major component of air pollution, and (4) water pollution measured as biochemical oxygen demand. Two measures of land use are examined: (1) number of national parks and other environmentally protected areas, and (2) forest area.[2]

International environmental institutions are measured in two ways. The first is nation-state ratification of international environmental treaties. Statistical analyses employ an index of the environmental treaties ratified by a country, based on thirteen major environmental treaties administered by the United Nations Environment Programme.[3] These treaties were selected because they are open to all countries and focus on environmental issues of broad importance in the international community.

The second institutional variable is national ties to international nongovernmental organizations (INGOs), which are the conventional way to operationalize the influence of world society and global culture.[4] INGO membership captures national ties to the international sphere, and is a broad indicator of environmental agents and penetration of international cultural discourses, as argued in Chapter 4. Boli and Thomas (1999) further argue that INGOs are the embodiment of global culture, and thus INGOs serve as a useful proxy to measure global cultural processes.

Statistical analyses must also control for factors that cause environmental degradation, including economic production (measured as the natural log of gross domestic product per capita) and a country's population size (logged population). Education levels (measured as secondary school enrollment per capita) are also included as a standard control variable.[5] Additional control variables (e.g., trade, democracy, civil war, and many others) did not change the results substantially. A brief set of control variables is presented here for the sake of simplicity.

[1] Data come from S. J. Smith et al. (2011).

[2] All dependent variables come from the World Development Indicators, World Bank (2013), except for data on sulfur dioxide as noted previously.

[3] The treaties include the Basel Convention, Convention on Biological Diversity, Convention on International Trade in Endangered Species of Wild Fauna and Flora, Convention on the Conservation of Migratory Species of Wild Animals, Kyoto Protocol, Ramsar Convention on Wetlands, Rotterdam Convention, Stockholm Convention on Persistent Organic Pollutants, UN Convention on the Law of the Sea, UN Convention to Combat Desertification, UN Framework Convention on Climate Change, Montreal Protocol, and the World Heritage Convention.

[4] INGOs are measured as the natural log of total INGOs in which citizens of a country hold membership. Data come from UIA (2010).

[5] Data for control variables come from the World Development Indicators, World Bank (2013).

Quantitative data are analyzed using panel regression models with fixed effects. This approach for analyzing cross-national longitudinal data is commonly used in sociology, political science, and economics. Panel models with fixed effects focus the analysis on longitudinal trends within cases, rather than comparing differences across cases. This approach holds historical differences constant across countries, avoiding problems due to the omission of time-invariant country-level variables. Rather than make apples-to-oranges comparisons of very different types of countries, this analysis tracks the individual apples over time. In doing so, fixed effects models avoid many kinds of omitted variable bias.

The models present the two institutional variables – environmental treaties and INGOs – separately, as they are two related facets of international institutions rather than independent phenomena. While these processes have been analytically distinguished here, case studies in this book suggest that they are deeply intertwined and frequently reinforce each other rather than having wholly independent causal effects.

Results

Tables A.1 and A.2 present statistical analyses that estimate the effects of institutional structures and agents on environmental conditions. Table A.1, shown below, indicates that institutional structures (environmental treaties) and environmental agents (INGOs) have a significant negative effect on four forms of pollution: carbon dioxide emissions (Model 1 and 2), sulfur dioxide (Model 3 and 4), nitrous oxides (Model 5 and 6), and water pollution (Model 7 and 8). These general effects are robust even when additional steps are taken to address potential sources of reverse causality (for instance, through dynamic models) and other possible confounding variables. These models also control for economic production (GDP per capita) and population (logged), which have significant positive effects on most of these forms of environmental pollution. In short, international institutions have beneficial effects, but other factors contribute to ongoing environmental damage.

Model 1 shows a statistically significant negative coefficient of environmental treaty ratification on CO_2 emissions. Likewise, the negative effect of INGOs in Model 2 indicates that countries with greater INGO linkages have reduced pollution emissions on average. These general associations may initially be non-intuitive, since most INGOs are not

TABLE A.1. *Fixed effects panel regression models: The effects of environmental treaties and international nongovernmental organizations (INGOs) on air pollution emissions and water quality, 1960–2010[a,b]*

VARIABLES	(1) CO_2	(2) CO_2	(3) SO_2	(4) SO_2	(5) NOx	(6) NOx	(7) Water BOD	(8) Water BOD
Population (log)	1.82*** (0.029)	1.41*** (0.031)	2.56*** (0.053)	2.13*** (0.064)	0.72*** (0.166)	0.48*** (0.111)	0.82*** (0.140)	0.36** (0.115)
GDP p/cap (log)	0.89*** (0.020)	0.69*** (0.021)	0.54*** (0.038)	0.26*** (0.045)	0.31*** (0.078)	0.32*** (0.062)	0.22*** (0.056)	-0.01 (0.041)
School enrollment	0.12* (0.047)	-0.07 (0.050)	-0.73*** (0.087)	-1.34*** (0.098)	0.20 (0.165)	0.23+ (0.134)	0.25** (0.095)	0.17+ (0.089)
Environmental treaties	-0.05*** (0.002)		-0.12*** (0.004)		-0.03*** (0.007)		-0.04*** (0.004)	
INGOs (log)		-0.03** (0.010)		-0.21*** (0.017)		-0.30*** (0.059)		-0.13*** (0.013)
Constant	-9.83*** (0.264)	-5.26*** (0.248)	-16.52*** (0.477)	-10.05*** (0.530)	1.19 (1.362)	4.68*** (0.760)	3.52** (1.149)	9.38*** (0.790)
Observations	4,478	4,112	3,475	3,186	412	393	844	836
R-squared	0.761	0.687	0.447	0.277	0.142	0.242	0.103	0.135
N of countries	176	175	123	121	126	123	91	90

Standard errors in parentheses.

*** $p<0.001$, ** $p<0.01$, * $p<0.05$.

[a] Time period of specific analyses varies depending on data availability.

[b] Negative coefficients indicate factors that reduce environmental pollution.

TABLE A.2. *Fixed effects panel regression models: The effects of environmental treaties and international nongovernmental organizations (INGOs) on protected area and forest area, 1990–2010[a]*

VARIABLES	(9)	(10)	(11)	(12)
	Protected area	Protected area	Forest area	Forest area
Population (log)	−2.18***	3.21***	−8.26***	−6.44***
	(0.554)	(0.495)	(0.516)	(0.372)
GDP p/cap (log)	0.08	1.74***	0.35	1.07***
	(0.260)	(0.250)	(0.242)	(0.192)
School enrollment	0.89	1.99***	−1.01*	−0.73
	(0.544)	(0.590)	(0.510)	(0.451)
Environmental treaties	0.38***		0.14***	
	(0.023)		(0.022)	
INGOs (log)		0.60***		0.39***
		(0.091)		(0.091)
Constant	21.44***	−30.95***	85.19***	65.15***
	(4.386)	(3.336)	(4.106)	(2.500)
Observations	2,456	2,344	2,412	2,299
R-squared	0.237	0.164	0.149	0.173
N of countries	176	173	177	174

Standard errors in parentheses.

*** $p<0.001$, ** $p<0.01$, * $p<0.05$.

[a] Positive coefficients indicate factors that improve environmental conditions.

oriented toward climate change issues. Indeed, it is the highly general nature of these findings that raises the central puzzle of the book: generalized ties to international institutions in world society matter, often as much or more than particular local policies. This finding supports the "Bee Swarm" model developed in Chapter 1.

Table A.2, shown above, turns to measures of land use, which relate to biodiversity, habitat loss, and deforestation. Table A.2 shows that institutional structures (treaties) and environmental agents (INGOs) have positive effects on the environmentally beneficial indicators of national parks and environmentally protected areas (Models 9 and 10) and forest area (Models 11 and 12). Again, economic size and population also have generally negative effects on these beneficial environmental measures, although economic size (GDP per capita) has a positive influence on forest area.

In sum: The analyses presented here provide an overview of the empirical literature that has shown that institutional structures and environmental agents reduce environmental degradation. These empirical results set up the primary puzzle explored in this book: What are the institutional mechanisms that bring about such changes in environmental conditions?

Bibliography

Amenta, Edwin. 2006. *When Movements Matter: The Townsend Plan and the Rise of Social Security.* Princeton: Princeton University Press.

Amenta, Edwin, and Kelly M. Ramsey. 2010. "Institutional Theory." In *The Handbook of Politics: State and Civil Society in Global Perspective,* edited by Kevin T. Leicht and J. Craig Jenkins, 15–39. New York: Springer.

Amenta, Edwin, and Neal Caren. 2004. "The Legislative, Organizational, and Beneficiary Consequences of State-Oriented Challengers." In *The Blackwell Companion to Social Movements,* edited by D. A. Snow, S. A. Soule, and H. Kriesi, 461–488. Malden: Blackwell.

Andersen, Mikael Skou, and Duncan Liefferink. 1997. "Introduction: The Impact of the Pioneers on EU Environmental Policy." In *European Environmental Policy: The Pioneers,* edited by Mikael Skou Andersen and Duncan Liefferink, 1–39. Manchester: Manchester University Park.

Andresen, Steinar. 1989. "Increased Public Attention: Communication and Polarization." In *International Resource Management: The Role of Science and Politics,* edited by Steinar Andresen and Willy Østreng, 25–45. New York: Belhaven Press.

Andresen, Steinar, Tora Skodvin, Arild Underdal, and Jørgen Wettestad. 2000. *Science and Politics in International Environmental Regimes: Between Integrity and Involvement.* Manchester: Manchester University Press.

Andersen, Stephen O., and K. Madhava Sarma. 2002. *Protecting the Ozone Layer: The United Nations History.* London: Earthscan Publications.

Andrews, Kenneth T., and Bob Edwards. 2005. "The Organizational Structure of Local Environmentalism." *Mobilization* 10 (2): 213–234.

Andrews, Richard N. L. 2006. *Managing the Environment, Managing Ourselves: A History of American Environmental Policy.* 2nd ed. New Haven: Yale University Press.

Arrow, Kenneth J. 1974. *The Limits of Organization.* New York: Norton.

Arts, Bas. 1998. *The Political Influence of Global NGOs: Case Studies on the Climate and Biodiversity Conventions.* Utrecht: International Books.

Baker, Randall. 1980. *Desertification: Cause and Control.* Norwich: School of Development Studies.

Barnett, Harold C. 1994. *Toxic Debts and the Superfund Dilemma.* Chapel Hill: University of North Carolina Press.

Barnett, Michael, and Martha Finnemore. 2004. *Rules for the World: International Organizations in Global Politics.* Ithaca: Cornell University Press.

Barratt, C. 1996. *Sustainable Development Case Study.* Teaching Politics Series. York: York University.

Barrett, Deborah, and Charles Kurzman. 2004. "Globalizing Social Movement Theory: The Case of Eugenics." *Theory and Society* 33 (5): 487–527.

Barrett, Deborah, and David John Frank. 1999. "Population Control for National Development: From World Discourse to National Policies." In *Constructing World Culture: International Nongovernmental Organizations since 1875*, edited by John Boli and George M. Thomas, 198–221. Stanford, CA: Stanford University Press.

Beck, Colin J. 2011. "The World-Cultural Origins of Revolutionary Waves: Five Centuries of European Contention." *Social Science History* 35 (2): 167–207.

Beck, Ulrich. 1999. *World Risk Society.* Malden: Blackwell.

Begley, Sharon. 2011. "Are You Ready for More?" *Newsweek*, May 29.

Benedick, Richard Elliot. 1998. *Ozone Diplomacy: New Directions in Safeguarding the Planet.* Cambridge, MA: Harvard University Press.

Berger, Peter L., and Thomas Luckmann. 1980[1967]. *The Social Construction of Reality: A Treatise in the Sociology of Knowledge.* New York: Irvington.

Bergesen, Albert J., and Laura Parisi. 1999. "Ecosociology and Toxic Emissions." In *Ecology and the World-System*, edited by Walter L. Goldfrank, David Goodman, and Andrew Szasz, 43–58. Westport: Greenwood Press.

Berkovitch, Nitza. 2002. *From Motherhood to Citizenship: Women's Rights and International Organizations.* Baltimore: Johns Hopkins University Press.

Bess, Michael. 2003. *The Light-Green Society: Ecology and Technological Modernity in France, 1960–2000.* Chicago: University of Chicago Press.

Betsill, Michele M. 2008. "Reflections on the Analytical Framework and NGO Diplomacy." In *NGO Diplomacy: The Influence of Nongovernmental Organizations in International Environmental Negotiations*, edited by Michele M. Betsill and Elisabeth Corell, 177–206. Cambridge, MA: MIT Press.

Blaikie, Piers, and John Mope Simo. 1998. "Cameroon's Environmental Accords: Signed, Sealed, but Undelivered." In *Engaging Countries: Strengthening Compliance with International Environmental Accords*, edited by Edith Brown Weiss and Harold K. Jacobson, 437–474. Cambridge, MA: MIT Press.

Bocking, Stephen. 2004. *Nature's Experts: Science, Politics, and the Environment.* New Brunswick: Rutgers University Press.

Boli, John. 2006. "The Rationalization of Virtue and Virtuosity in World Society." In *Transnational Governance: Institutional Dynamics of Regulation*, edited by Marie-Laure Djelic and Kerstin Sahlin-Andersson, 95–118. Cambridge: Cambridge University Press.

Boli, John, Francisco O. Ramirez, and John W. Meyer. 1985. "Explaining the Origins and Expansion of Mass Education." *Comparative Education Review* 29 (2): 145–168.

Boli, John, and George M. Thomas. 1997. "World Culture in the World Polity: A Century of International Non-governmental Organization." *American Sociological Review* 62 (2): 171–190.

——— eds. 1999. *Constructing World Culture: International Nongovernmental Organizations since 1875.* Stanford, CA: Stanford University Press.

Boyle, Elizabeth Heger. 2002. *Female Genital Cutting: Cultural Conflict in the Global Community.* Baltimore: Johns Hopkins Press.

Boyle, Elizabeth Heger, and Sharon E. Preves. 2000. "National Politics as International Process: The Case of Anti-Female-Genital-Cutting Laws." *Law & Society Review* 34 (3): 703–737.

Bramble, Barbara J., and Gareth Porter. 1992. "Non-governmental Organizations and the Making of U.S. International Environmental Policy." In *The International Politics of the Environment: Actors, Interests, and Institutions,* edited by Andrew Hurrell and Benedict Kingsbury, 313–353. Oxford: Clarendon Press.

Brand, Karl-Werner. 1999. "Dialectics of Institutionalisation: The Transformation of the Environmental Movement in Germany." In *Environmental Movements: Local, National and Global,* edited by Christopher Rootes, 35–58. Portland: Frank Cass.

Breitmeier, Helmut, and Volker Rittberger. 2000. "Environmental NGOs in an Emerging Global Civil Society." In *The Global Environment in the Twenty-First Century: Prospects for International Cooperation,* edited by Pamela S. Chasek, 130–163. New York: United Nations University Press.

Brenton, Tony. 1994. *The Greening of Machiavelli: The Evolution of International Environmental Politics.* London: Earthscan Publications.

Brimblecombe, Peter. 1987. *The Big Smoke: A History of Air Pollution in London since Medieval Times.* London: Methuen.

Broadbent, Jeffrey. 1998. *Environmental Politics in Japan: Networks of Power and Protest.* Cambridge: Cambridge University Press.

Bromley, Patricia, John W. Meyer, and Francisco O. Ramirez. 2011. "The Worldwide Spread of Environmental Discourse in Social Studies, History, and Civics Textbooks, 1970–2008." *Comparative Education Review* 55 (4): 517–545.

Bromley, Patricia, and W. W. Powell. 2012. "From Smoke and Mirrors to Walking the Talk: Decoupling in the Contemporary World." *Academy of Management Annals* 6:1–48.

Brown, J. Christopher. 2006. "Placing Local Environmental Protest within Global Environmental Networks: Colonist Farmers and Sustainable Development in the Brazilian Amazon." In *Shades of Green: Environmental Activism around the Globe,* edited by Christof Mauch, Nathan Stoltzfus, and Douglas R. Weiner, 197–218. Lanham: Rowman and Littlefield.

Brown, Katrina, and David W. Pearce. 1994. *The Causes of Tropical Deforestation.* London: UCL Press.

Brown Weiss, Edith. 1998. "The Five International Treaties: A Living History." In *Engaging Countries: Strengthening Compliance with International Environmental Accords*, edited by Edith Brown Weiss and Harold K. Jacobson, 89–172. Cambridge, MA: MIT Press.

Brunnée, Jutta. 1988. *Acid Rain and Ozone Layer Depletion: International Law and Regulation*. Dobbs Ferry: Transnational Publishers.

Bryant, Raymond L. 2005. *Nongovernmental Organizations in Environmental Struggles: Politics and the Makings of Moral Capital in the Philippines*. New Haven: Yale University Press.

Bunker, Stephen G. 1985. *Underdeveloping the Amazon: Extraction, Unequal Exchange, and the Failure of the Modern State*. Urbana: University of Illinois Press.

Bunker, Stephen G., and Paul S. Ciccantell. 1999. "Economic Ascent and the Global Environment: World-Systems Theory and the New Historical Materialism." In *Ecology and the World-System*, edited by Walter L. Goldfrank, David Goodman, and Andrew Szasz, 107–122. Westport: Greenwood Press.

Burchell, Jon. 2002. *The Evolution of Green Politics: Development and Change within European Green Parties*. Sterling: Earthscan Publications.

Burstein, Paul. 1999. "Social Movements and Public Policy." In *How Social Movements Matter*, edited by Marco Giugni, Doug McAdam, and Charles Tilly, 3–21. Minneapolis: University of Minnesota Press.

Burt, Ronald S. 1992. *Structural Holes: The Social Structure of Competition*. Cambridge, MA: Harvard University Press.

Buttel, Frederick H. 2000. "World Society, the Nation-State, and Environmental Protection: Comment on Frank, Hironaka, and Schofer." *American Sociological Review* 65:117–121.

Cadman, Timothy. 2011. *Quality and Legitimacy of Global Governance: Case Lessons from Forestry*. New York: Palgrave Macmillan.

Caldwell, Lynton Keith. 1984. *International Environmental Policy: Emergence and Dimensions*. Durham: Duke University Press.

1990. *Between Two Worlds: Science, the Environmental Movement, and Policy Choice*. Cambridge: Cambridge University Press.

California Energy Commission. 2006. "Inventory of California Greenhouse Gas Emissions and Sinks: 1990–2004." CEC-600-2006-013-SF.

Carruthers, Bruce G., and Sarah L. Babb. 2012. *Economy/Society: Markets, Meanings, and Social Structure*. Los Angeles: Sage.

Castilla, Emilio J. 2009. "The Institutional Production of National Science in the 20th Century." *International Sociology* 24 (6): 833–869.

Catton, William R., Jr. 1982. *Overshoot: The Ecological Basis of Revolutionary Change*. Urbana-Champaign: University of Illinois Press.

Chabbott, Colette. 2003. *Constructing Education for Development: International Organizations and Education for All*. New York: RoutledgeFalmer.

Chase-Dunn, Christopher 1989. *Global Formation: Structures of the World-Economy*. Cambridge: Basil Blackwell.

Chatterjee, Pratap, and Matthias Finger 1994. *The Earth Brokers: Power, Politics, and World Development*. London: Routledge.

Chayes, Abram, and Antonia Handler Chayes. 1995. *The New Sovereignty: Compliance with International Regulatory Agreements*. Cambridge, MA: Harvard University Press.

2001[1993]. "On Compliance." In *International Institutions: An International Organization Reader*, edited by Lisa L. Martin and Beth A. Simmons, 247–277. Cambridge, MA: MIT Press.

Chirot, Daniel. 1986. *Social Change in the Modern Era*. New York: Harcourt.

Clapp, Jennifer. 2001. *Toxic Exports: The Transfer of Hazardous Wastes from Rich to Poor Countries*. Ithaca: Cornell University Press.

CNN Money. 2010. "Best Jobs in America." Accessed September 21, 2011. http://money.cnn.com/magazines/moneymag/bestjobs/2010/jobgrowth/index.html

Cole, Wade M. 2005. "Sovereignty Relinquished? Explaining Commitment to the International Human Rights Covenants, 1966–1999." *American Sociological Review* 70 (3): 472–495.

2011. *Uncommon Schools: The Global Rise of Postsecondary Institutions for Indigenous Peoples*. Stanford, CA: Stanford University Press.

Cole, Wade M., and Francisco O. Ramirez. 2013. "Conditional Decoupling: Assessing the Impact of National Human Rights Institutions, 1981 to 2004." *American Sociological Review* 78 (4): 702–725.

Commoner, Barry. 1971. *The Closing Circle: Nature, Man, and Technology*. New York: Alfred A. Knopf.

Cooley, Charles Horton. 1998. *On Self and Social Organization*. Chicago: University of Chicago Press.

Crenson, Matthew A. 1971. *The Un-politics of Air Pollution: A Study of Non-decision Making in the Cities*. Baltimore: Johns Hopkins University Press.

Della Porta, Donatella, and Mario Diani. 2006. *Social Movements: An Introduction*. Malden: Blackwell.

de Lupis, Ingrid Detter. 1989. "The Human Environment: Stockholm and Its Follow Up." In *Global Issues in the United Nations' Framework*, edited by Paul Taylor and A. J. R. Groom, 205–225. London: Macmillan Press.

DeSombre, Elizabeth R. 2007. *The Global Environment and World Politics*. 2nd ed. New York: Continuum International.

DiMaggio, Paul J. 1988. "Interest and Agency in Institutional Theory." In *Institutional Patterns and Organizations: Culture and Environment*, edited by Lynne G. Zucker, 3–22. Cambridge: Ballinger.

DiMaggio, Paul J., and W. W. Powell. 1983. "The Iron Cage Revisited: Institutional Isomorphism and Collective Rationality in Organizational Fields." *American Sociological Review* 48 (2): 147–160.

Dimitrov, Radoslav S. 2006. *Science and International Environmental Policy: Regimes and Nonregimes in Global Governance*. Lanham: Rowman and Littlefield.

Dobbin, Frank. 1994. *Forging Industrial Policy: The United States, Britain, and France in the Railway Age*. New York: Cambridge University Press.

2009. *Inventing Equal Opportunity*. Princeton: Princeton University Press.

Doherty, Brian. 2002. *Ideas and Actions in the Green Movement*. New York: Routledge.

Downs, Anthony. 1972. "Up and Down with Ecology: The Issue Attention Cycle." *Public Interest* **28**:38–50.

Downs, George W., David M. Rocke, and Peter N. Barsoom. 2001[1993] "Is the Good News about Compliance Good News about Cooperation?" In *International Institutions: An International Organization Reader*, edited by Lisa L. Martin and Beth A. Simmons, 279–306. Cambridge, MA: MIT Press.

Drori, Gili S., John W. Meyer, and Hokyu Hwang, eds. 2006. *Globalization and Organization: World Society and Organizational Change*. Oxford: Oxford University Press.

Drori, Gili S., John W. Meyer, Francisco O. Ramirez, and Evan Schofer. 2003. *Science in the Modern World Polity: Institutionalization and Globalization*. Stanford: Stanford University Press.

Dunlap, Riley E. 1995. "Environmental Concerns and the Third World." *Science* **268** (5214):1114–1115.

Dunlap, Riley E., and Angela G. Mertig, eds. 1992. *American Environmentalism: The U.S. Environmental Movement, 1970–1990*. Philadelphia: Taylor and Francis.

Dunlap, Riley E., and Richard York. 2008. "The Globalization of Environmental Concern and the Limits of the Postmaterialist Values Explanation: Evidence from Four Multinational Surveys." *The Sociological Quarterly* **49** (3): 529–563.

Ecolex. 2010. Accessed May 5, 2010. www.ecolex.org

Ehrlich, Paul. 1968. *The Population Bomb*. New York: Ballantine.

El-Hinawi, Essam, and Manzur-Ul-Haque Hashmi, eds. 1982. *Global Environmental Issues: United Nations Environment Programme*. Dublin: Tycooly International Pub.

Espeland, Wendy Nelson, and Mitchell L. Stevens. 1998. "Commensuration as a Social Process." *Annual Review of Sociology* **24**: 313–343.

Evans, Peter B., Harold K. Jacobson, and Robert D. Putnam, eds. 1993. *Double-Edged Diplomacy: International Bargaining and Domestic Politics*. Berkeley: University of California Press.

Falkner, Robert. 2008. *Business Power and Conflict in International Environmental Politics*. New York: Palgrave Macmillan.

2010. "Business and Global Climate Governance: A Neo-pluralist Perspective." In *Business and Global Governance*, edited by Morten Ougaard and Anna Leander, 99–117. New York: Routledge.

Ferguson, James. 1990. *The Anti-politics Machine: "Development," Depoliticization and Bureaucratic Power in Lesotho*. Minneapolis: University of Minnesota Press.

Ferree, Myra Marx, William A. Gamson, Jürgen Gerhards, and Dieter Ruch. 2002. *Shaping Abortion Discourse: Democracy and the Public Sphere in Germany and the United States*. Cambridge: Cambridge University Press.

Fisher, Dana R. 2003. "Global and Domestic Actors within the Global Climate Change Regime: Toward a Theory of the Global Environmental System." *International Journal of Sociology and Social Policy* **23** (10): 5–30.

2004. *National Governance and the Global Climate Change Regime*. Lanham: Rowman and Littlefield.

2013. "Understanding the Relationship between Sub-National and National Climate Change Politics in the United States: Toward a Theory of Boomerang Federalism." *Environment & Planning C: Government & Policy* 31(5): 769–784.

Fisher, Dana R., and William R. Freudenburg. 2001. "Ecological Modernization and Its Critics: Assessing the Past and Looking toward the Future." *Society and Natural Resources* 14 (8): 701–709.

2002. "Postindustrialization and Environmental Quality: An Empirical Analysis of the Environmental State." *Social Forces* 83 (1): 157–188.

Fisher, Dana R., Joseph Waggle, and Philip Leifeld. 2013. "Where Does Political Polarization Come From? Locating Polarization with the U.S. Climate Change Debate." *American Behavioral Scientist.* 57 (1): 70–92.

Fligstein, Neil. 1997. "Social Skill and Institutional Theory." *American Behavioral Scientist* 40 (4): 397–405.

2001. "Social Skill and the Theory of Fields." *Sociological Theory* 19:105–125.

Foucault, Michel. 1990 *History of Sexuality, Vol I*, translated by Robert Hurley. New York: Vintage.

Fourcade, Marion. 2011. "Cents and Sensibility: Economic Valuation and the Nature of 'Nature.'" *American Journal of Sociology* 116 (6): 1721–1777.

Fox, Jonathan A., and L. David Brown. 1998. "Introduction." In *The Struggle for Accountability*, edited by Jonathan A. Fox and L. David Brown, 1–48. Cambridge, MA: MIT Press.

Frank, David John. 1997. "Science, Nature, and the Globalization of the Environment, 1870–1990." *Social Forces* 76 (2): 409–435.

1999. "The Social Bases of Environmental Treaty Ratification, 1900–1960." *Sociological Inquiry* 69 (4): 523–550.

Frank, David John, and Jay Gabler. 2006. *Reconstructing the University: Worldwide Shifts in Academia in the 20th Century*. Stanford, CA: Stanford University Press.

Frank, David John, Tara Hardinge, and Kassia Wosick Correa. 2009. "The Global Dimensions of Rape-Law Reform: A Cross National Study of Policy Outcomes." *American Sociological Review* 74 (2): 272–290.

Frank, David John, Ann Hironaka, John W. Meyer, Evan Schofer, and Nancy Tuma. 1999. "The Rationalization and Organization of Nature in World Culture." In *Constructing World Culture: International Nongovernmental Organizations since 1875*, edited by John Boli and George M. Thomas, 81–99. Stanford, CA: Stanford University Press.

Frank, David John, Ann Hironaka, and Evan Schofer. 2000a. "The Nation-State and the Natural Environment over the Twentieth Century." *American Sociological Review* 65 (1): 96–116.

2000b. "Environmentalism as a Global Institution: Reply to Buttel." *American Sociological Review* 65 (1): 122–127.

Frank, David John, Wesley Longhofer, and Evan Schofer. 2007. "World Society, NGOs, and Environmental Policy Reform in Asia." *International Journal of Comparative Sociology* 48: 275–295.

Frank, David John, and John W. Meyer. 2002. "The Profusion of Individual Roles and Identities in the Post War Period." *Sociological Theory* 20 (1): 86–105.

2007. "University Expansion and the Knowledge Society." *Theory and Society* 36 (4): 287–311.

Frank, David John, Karen Jeong Robinson, and Jared Olesen. 2011. "The Global Expansion of Environmental Education in Universities." *Comparative Education Review* 55 (November): 546–573.

Freudenburg, William R., and Robert Gramling. 1994. *Oil in Troubled Waters: Perceptions, Politics, and the Battle over Offshore Drilling.* Albany: State University Press of New York.

Gamson, William. 1990. *The Strategy of Social Protest.* Homewood: Dorsey Press.

Gellert, Paul K. 2010. "Rival Transnational Networks, Domestic Politics, and Indonesian Timber." *Journal of Contemporary Asia* 40 (4): 539–567.

Giddens, Anthony. 2009. *The Politics of Climate Change.* Malden: Polity.

Giugni, Marco. 2004. *Social Protest and Policy Change: Ecology, Antinuclear, and Peace Movements in Comparative Perspective.* Lanham: Rowman and Littlefield.

Glennon, Michael J., and Alison L. Stewart. 1998. "The United States: Taking Environmental Treaties Seriously." In *Engaging Countries: Strengthening Compliance with International Environmental Accords,* edited by Edith Brown Weiss and Harold K. Jacobson, 173–214. Cambridge, MA: MIT Press.

Goertz, Gary. 2003. *International Norms and Decision Making: A Punctuated Equilibrium Model.* Lanham: Rowman and Littlefield.

Goffman, Erving. 1959. *The Presentation of Self in Everyday Life.* New York: Anchor Books.

1974. *Frame Analysis: An Essay on the Organization of Experience.* Cambridge, MA: Harvard University Press.

Goldman, Michael. 2005. *Imperial Nature: The World Bank and Struggles for Social Justice in the Age of Globalization.* New Haven: Yale University Press.

Gould, Kenneth A., Allan Schnaiberg, and Adam S. Weinberg. 1996. *Local Environmental Struggles: Citizen Activism in the Treadmill of Production.* Cambridge: Cambridge University Press.

Gould, Kenneth A., David N. Pellow, and Allan Schnaiberg. 2008. *The Treadmill of Production: Injustice and Unsustainability in the Global Economy.* Boulder: Paradigm.

Grainger, Alan. 1990. *The Threatening Desert: Controlling Desertification.* London: Earthscan.

Granovetter, Mark S. 1973. "The Strength of Weak Ties." *American Journal of Sociology* 78 (6): 1360–1380.

Grant, Don Sherman II, Albert J. Bergesen, and Andrew W. Jones. 2002. "Organizational Size and Pollution: The Case of the U.S. Chemical Industry." *American Sociological Review* 67 (3): 389–407.

Guimarães, Roberto P. 1991. *The Ecopolitics of Development in the Third World: Politics and Environment in Brazil.* Boulder: Lynne Rienner.

Guruswamy, Lakshman. 2007. *International Environmental Law in a Nutshell.* 3rd ed. St. Paul, MN: Thomson/West.

Gusfield, Joseph R. 1981. *The Culture of Public Problems: Drinking-Driving and the Symbolic Order*. Chicago: University of Chicago Press.

Haas, Ernst B., Mary Pat Williams, and Don Babai. 1977. *Scientists and World Order: The Uses of Technical Knowledge in International Organizations*. Berkeley: University of California Press.

Haas, Peter M. 1992. "Banning Chlorofluorocarbons: Epistemic Community Efforts to Protect Stratospheric Ozone." *International Organization* 46 (1): 187–224.

Hadler, Markus, and Max Haller. 2011. "Global Activism and Nationally Driven Recycling: The Influence of World Society and National Contexts on Public and Private Environmental Behavior." *International Sociology* 26 (3): 315–345.

Hafner-Burton, Emily, and Kiyoteru Tsutsui. 2005. "Human Rights Practices in a Globalizing World: The Paradox of Empty Promises." *American Journal of Sociology* 110 (5): 1373–1411.

Hall, Peter A. 1989. *The Political Power of Economic Ideas: Keynesianism across Nations*. Princeton: Princeton University Press.

Hallett, Tim, and Marc J. Ventresca. 2006. "Inhabited Institutions: Social Interactions and Organizational Forms in Gouldner's Patterns of Industrial Bureaucracy." *Theory and Society* 35 (2): 213–236.

Hamilton, James T. 2005. *Regulation through Revelation: The Origin, Politics, and Impacts of the Toxics Release Inventory Program*. Cambridge: Cambridge University Press.

Harding, Garrett. 1968. "The Tragedy of the Commons." *Science* 162:1243–1248.

Harris, Paul G., ed. 2000, *Climate Change and American Foreign Policy*. New York, NY: Palgrave Macmillan.

Harrison, Neil E., and Gary C. Bryner, eds. 2004. *Science and Politics in the International Environment*. Lanham: Rowman and Littlefield.

Hayden, Sherman Strong. 1970[1942]. *The International Protection of Wild Life: An Examination of Treaties and Other Agreements for the Preservation of Birds and Mammals*. New York: AMS Press.

Hayes, Samuel P. 1959. *Conservation and the Gospel of Efficiency*. Cambridge, MA: Harvard University Press.

Hecht, Susanna, and Alexander Cockburn. 1989. *The Fate of the Earth: Developers, Destroyers, and Defenders of the Amazon*. New York: Verso.

Hechter, Michael, and Karl-Dieter Opp, eds. 2001. *Social Norms*. New York: Russell Sage Foundation.

Heimer, Carol A. 1999. "Competing Institutions: Law, Medicine, and Family in Neonatal Intensive Care. *Law and Society Review*. 33 (1): 17–66.

Herring, Ronald J., and Erach Bharucha. 1998. "India: Embedded Capacities." In *Engaging Countries: Strengthening Compliance with International Environmental Accords*, edited by Edith Brown Weiss and Harold K. Jacobson, 395–436. Cambridge: MIT Press.

Hicks, Alexander. 1999. *Social Democracy and Welfare Capitalism: A Century of Income Security Politics*. Ithaca: Cornell University Press.

Hironaka, Ann. 2002. "The Globalization of Environmental Protection: The Case of Environmental Impact Assessment." *International Journal of Comparative Sociology* 43 (1): 65–78.

2003. "Science and the Environment." In *Science in the Modern World Polity: Institutionalization and Globalization*, edited by Gili S. Drori, John W. Meyer, Francisco O. Ramirez, and Evan Schofer, 484–515. Stanford, CA: Stanford University Press.

2005. *Neverending Wars: Weak States, the International Community, and the Perpetuation of Civil War*. Cambridge, MA: Harvard University Press.

Hironaka, Ann, and Evan Schofer. 2002. "Decoupling in the Environmental Arena: The Case of the Environmental Impact Assessment." In *Organizations, Policy, and the Natural Environment: Institutional and Strategic Perspectives*, edited by Andrew J. Hoffman and Marc J. Ventresca, 214–231. Stanford, CA: Stanford University Press.

Hobsbawm, Eric J. 1964. *The Age of Revolution, 1789–1848*. New York: New American Library.

Hoffman, Andrew J. 2000. *Competitive Environmental Strategy*. Washington, DC: Island Press.

2001. *From Heresy to Dogma: An Institutional History of Corporate Environmentalism*. Stanford: Stanford University Press.

2007. *Carbon Strategies: How Leading Companies Are Reducing Their Climate Change Footprint*. Ann Arbor: University of Michigan Press.

Hoffman, Matthew J. 2005. *Ozone Depletion and Climate Change: Constructing a Global Response*. Albany: State University of New York Press.

Humphreys, David. 2006. *Logjam: Deforestation and the Crisis of Global Governance*. Sterling: Earthscan.

Hurrell, Andrew. 1992. "Brazil and the International Politics of Amazonian Deforestation." In *The International Politics of the Environment: Actors, Interests, and Institutions*, edited by Andrew Hurrell and Benedict Kingsbury, 398–429. Oxford: Clarendon Press.

Hurrell, Andrew, and Benedict Kingsbury. 1992. "The International Politics of the Environment: An Introduction." In *The International Politics of the Environment: Actors, Interests, and Institutions*, edited by Andrew Hurrell and Benedict Kingsbury, 1–50. Oxford: Clarendon Press.

Hutton, Drew, and Libby Connors 1999. *A History of the Australian Environment Movement*. Cambridge: Cambridge University Press.

Ignatow, Gabriel. 2007. *Transnational Identity Politics and the Environment*. Lanham: Lexington Books.

Inglehart, Ronald. 1977. *The Silent Revolution: Changing Values and Political Styles among Western Publics*. Princeton: Princeton University Press.

1990. *Culture Shift in Advanced Industrial Society*. Princeton: Princeton University Press.

Intergovernmental Panel on Climate Change (IPCC). 2007. *Climate Change 2007: Synthesis Report*. Geneva: IPCC.

Jachtenfuchs, Markus. 1990. "The European Community and the Protection of the Ozone Layer." *Journal of Common Market Studies*. 28 (3): 261–277.

Jacques, Peter J., Riley E. Dunlap, and Mark Freeman. 2008. "The Organisation of Denial: Conservative Think Tanks and Environmental Scepticism." *Environmental Politics* 17 (3): 349–385.

Jang, Yong Suk. 2000. "The Worldwide Formation of Ministries of Science and Technology, 1950–1990." *Sociological Perspectives* 43 (2): 247–270.

Jänicke, Martin. 1990. *State Failure: The Impotence of Politics in Industrial Society*. Oxford: Polity Press.

———. 1996. "The Political System's Capacity for Environmental Policy." In *National Environmental Policies: A Comparative Study of Capacity-Building*, edited by Martin Jänicke and Helmut Weidner, 1–24. New York: Springer-Verlag.

Jänicke, Martin, and Helmut Weidner, eds. 1995. *Successful Environmental Policy: A Critical Evaluation of 24 Cases*. Berlin: Edition Sigma.

———. 1996. "Summary: Global Environmental Policy Learning." In *National Environmental Policies: A Comparative Study of Capacity-Building*, edited by Martin Jänicke and Helmut Weidner, 299–314. New York: Springer-Verlag.

Jasanoff, Sheila. 1990. *The Fifth Branch: Science Advisers as Policymakers*. Cambridge, MA: Harvard University Press.

Jenkins, J. Craig, and William Form. 2005. "Social Movements and Social Change." In *The Handbook of Political Sociology: States, Civil Societies, and Globalization*, edited by T. Janoski, R. Alford, A. Hicks, and M. A. Schwartz, chapter 15. New York: Cambridge University Press.

Jepperson, Ronald L. 1991. "Institutions, Institutional Effects, and Institutionalism." In *The New Institutionalism in Organizational Analysis*, edited by Walter W. Powell and Paul J. DiMaggio, 143–163. Chicago: University of Chicago Press.

Johnson, Erik, and John D. McCarthy. 2005. "The Sequencing of Transnational and National Social Movement Mobilization: The Organizational Mobilization of the Global and U.S. Environmental Movements." In *Transnational Protest and Global Activism*, edited by Donatella della Porta and Sidney Tarrow, 71–94. Lanham: Rowman and Littlefield.

Jones, Charles A., and David L. Levy. 2009. "Business Strategies and Climate Change." In *Changing Climates in North American Politics*, edited by Henrik Selin and Stacy D. VanDeveer, 219–240. Cambridge, MA: MIT Press.

Jorgenson, Andrew K. 2003. "Consumption and Environmental Degradation: A Cross-National Analysis of the Ecological Footprint. *Social Problems* 50 (3): 374–394.

———. 2006. "Global Warming and the Neglected Greenhouse Gas: A Cross-National Study of the Social Causes of Methane Emissions Intensity, 1995." *Social Forces* 84 (3): 1779–1798.

———. 2007. "Does Foreign Investment Harm the Air We Breathe and the Water We Drink? A Cross-National Study of Carbon Dioxide Emissions and Organic Water Pollution in Less-Developed Countries, 1975–2000." *Organization and Environment* 20 (2): 137–156.

Kamens, David, and Aaron Benavot. 1991. "Elite Knowledge for the Masses: The Origins and Spread of Mathematics and Science Education in National Curricula." *American Journal of Education* 99 (2): 137–180.

Kamieniecki, Sheldon. 2006. *Corporate America and Environmental Policy: How Often Does Business Get Its Way?* Stanford: Stanford University Press.

Keck, Margaret E., and Kathryn Sikkink. 1998. *Activists beyond Borders*. Ithaca: Cornell University Press.

Keohane, Robert O., and Marc A. Levy, eds. 1996. *Institutions for Environmental Aid: Pitfalls and Promises*. Cambridge, MA: MIT Press.

Keohane, Robert O., Peter M. Haas, and Marc A. Levy. 1993. "The Effectiveness of International Environmental Institutions." In *Institutions for the Earth: Sources of Effective International Environmental Protection*, edited by Peter M. Haas, Robert O. Keohane, and Marc A. Levy, 3–26. Cambridge, MA: MIT Press.

Khagram, Sanjeev, James V. Riker, and Kathryn Sikkink. 2002. "From Santiago to Seattle: Transnational Advocacy Groups Restructuring World Politics." In *Restructuring World Politics: Transnational Social Movements, Networks, and Norms*, edited by Sanjeev Khagram, James V. Riker, and Kathryn Sikkink, 3–23. Minneapolis: University of Minnesota Press.

Khator, Renu. 1991. *Environment, Development, and Politics in India*. Lanham: University Press of America.

Kitschelt, Herbert. 1993. "The Green Phenomenon in Western Party Systems." In *Environmental Politics in the International Arena: Movements, Parties, Organizations, and Policy*, edited by Sheldon Kamieniecki, 93–112. Albany: State University of New York Press.

Kluger, Jeffrey. "Earth at the Tipping Point." *Time*, March 26, 2006. Accessed September 18, 2011. http://www.time.com/time/magazine/article/0,9171,1176980,00.html.

Kraft, Michael E., and Ruth Kraut. 1988. "Citizen Participation and Hazardous Waste Policy Implementation." In *Dimensions of Hazardous Waste Politics and Policy*, edited by Charles E. Davis and James P. Lester, 63–80. New York: Greenwood Press.

Krane, Dale. 2007. "The Middle Tier in American Federalism: State Government Policy Activism during the Bush Presidency." *Publius: The Journal of Federalism* 37:453–477.

Kummer, Katharina. 1994. *Transboundary Movements of Hazardous Wastes at the Interface of Environment and Trade*. Geneva: UNEP

Kymlicka, Will. 1996. *Multicultural Citizenship: A Liberal Theory of Minority Rights*. Oxford: Oxford University Press.

Lamont, Michele, and Mario Luis Small. 2008. "How Culture Matters: Enriching Our Understanding of Poverty." In *The Colors of Poverty: Why Racial and Ethnic Disparities Persist*, edited by David Harris and Ann Lin, 76–102. New York: Russell Sage.

Leonard, H. Jeffrey. 1988. *Pollution and the Struggle for the World Product: Multinational Corporations, Environment, and International Comparative Advantage*. Cambridge: Cambridge University Press.

Le Prestre, Philippe. 1989. *The World Bank and the Environmental Challenge*. Selinsgrove: Susquehanna University Press.

Levy, Marc A., Robert O. Keohane, and Peter M. Haas. 1993. "Improving the Effectiveness of International Environmental Institutions." In *Institutions for the Earth: Sources of Effective International Environmental Protection*, edited by Peter M. Haas, Robert O. Keohane and Marc A. Levy, 397–426. Cambridge, MA: MIT Press.

Lidskog, Rolf, and Göran Sundqvist. 2002. "The Role of Science in Environmental Regimes: The Case of LRTAP." *European Journal of International Relations* 8 (1): 77–101.

Lim, Alwyn, and Kiyoteru Tsutsui 2011. "Globalization and Commitment in Corporate Social Responsibility: Cross-National Analyses of Institutional and Political-Economy Effects." *American Sociological Review* 77 (1): 69–98.

Lindstrom, Matthew J., and Zachary A. Smith. 2001. *The National Environmental Policy Act: Judicial Misconstruction, Legislative Indifference, and Executive Neglect.* College Station: Texas A&M Press.

Litfin, Karen T. 1994. *Ozone Discourses: Science and Politics in Global Environmental Cooperation.* New York: Columbia University Press.

Longhofer, Wesley, and Evan Schofer. 2010. "National and Global Origins of Environmental Association." *American Sociological Review* 71 (4): 505–533.

Lounsbury, Michael, and Ellen T. Crumley. 2007. "New Practice Creation: An Institutional Perspective on Innovation." *Organization Studies* 28 (7): 993–1012.

Lukes, Steven. 1974. *Power: A Radical View.* New York: Macmillan.

Lutsey, Nicholas, and Daniel Sperling. 2008. "America's Bottom-Up Climate Change Mitigation Policy." *Energy Policy* 36:673–685.

Mahoney, James, and Kathleen Thelen. 2010. "A Theory of Gradual Institutional Change." In *Explaining Institutional Change: Ambiguity, Agency, and Power*, edited by James Mahoney and Kathleen Thelen, 1–37, Cambridge: Cambridge University Press.

Mangun, William R. 1988. "A Comparative Analysis of Hazardous Waste Management Policy in Western Europe." In *Dimensions of Hazardous Waste Politics and Policy*, edited by Charles E. Davis and James P. Lester, 205–222. New York: Greenwood Press.

March, James G. 1981. "Footnotes to Organizational Change." *Administrative Science Quarterly* 26 (4): 563–577.

March, James G., and Johan P. Olsen. 1984. The New Institutionalism: Organizational Factors in Political Life." *American Political Science Review* 78:734–749.

Mazmanian, Daniel A., and David Morell. 1992. *Beyond Superfailure: America's Toxics Policy for the 1990s.* Boulder: Westview Press.

Mazmanian, Daniel A., and Jeanne Nienaber. 1979. *Can Organizations Change: Environmental Protection, Citizen Participation, and the Corps of Engineers.* Washington, DC: The Brookings Institute.

Mazmanian, Daniel A., and Paul A. Sabatier. 1983. *Implementation and Public Policy.* Glenview: Scott, Foresman and Company.

McCarthy, John D., and Mayer N. Zald. 1977. "Resource Mobilization and Social Movements: A Partial Theory." *American Journal of Sociology* 82 (6): 1212–1241.

McCormick, John. 1995. *The Global Environmental Movement.* 2nd ed. Chichester: John Wiley and Sons.

1999. "The Role of Environmental NGOs in International Regimes." In *The Global Environment: Institutions, Law, and Policy*, edited by Norman J. Vig and Regina S. Axelrod, 52–72. Washington, DC: Congressional Quarterly Press.

Mead, George Herbert. 1934. *Mind, Self, and Society*. Chicago: University of Chicago Press.

Meyer, David S., and Deana A. Rohlinger. 2012. "Big Books and Social Movements: A Myth of Ideas and Social Change." *Social Problems* **59** (1): 136–159.

Meyer, David S., and Suzanne Staggenborg. 1996. "Movements, Countermovements and the Structure of Political Opportunity." *American Journal of Sociology* **101** (6): 1628–1660.

Meyer, John W. 1980. "The World Polity and the Authority of the Nation-State." In *Studies of the Modern World-System*, edited by Albert J. Bergesen, 109–137. New York: Academic Press.

2002. "Forward." In *Organizations, Policy, and the Natural Environment*, edited by Andrew J. Hoffman and Marc J. Ventresca, xiii-xvii. Stanford: Stanford University Press.

2010. "World Society, Institutional Theories, and the Actor." *Annual Review of Sociology* **36**:1–20.

Meyer, John W., John Boli, George M. Thomas, and Francisco O. Ramirez. 1997a. "World Society and the Nation-State." *American Journal of Sociology* **103** (1): 144–181.

Meyer, John W., David John Frank, Ann Hironaka, Evan Schofer, and Nancy Brandon Tuma. 1997b. "The Structuring of a World Environmental Regime, 1870–1990." *International Organization* **51** (4): 623–651.

Meyer, John W., and Michael T. Hannan, eds. 1979. *National Development and the World System: Educational, Economic, and Political Change, 1950–1970*. Chicago: University of Chicago Press.

Meyer, John W., and Ronald L. Jepperson. 2000. "The 'Actors' of Modern Society: The Cultural Construction of Social Agency." *Sociological Theory* **18** (1): 100–120.

Meyer, John W., and Brian Rowan. 1977. "Institutionalized Organizations: Formal Structure as Myth and Ceremony." *American Journal of Sociology* **83** (2): 340–363.

Meyer, John W., Francisco O. Ramirez, and Yasemin Nuhoglu Soysal. 1992. "World Expansion of Mass Education, 1870–1970." *Sociology of Education* **65** (2): 128–49.

Miller, Clark A., and Paul N. Edwards, eds. 2001. *Changing the Atmosphere: Expert Knowledge and Environmental Governance*. Cambridge, MA: MIT Press.

Mizruchi, Mark, and Joseph Galaskiewicz. 1993. "Networks of Interorganization Relations." *Sociological Methods and Research* **22**:46–70.

Mol, Arthur P. J. 1997. "Ecological Modernization: Industrial Transformations and Environmental Reform." In *The International Handbook of Environmental Sociology*, edited by Michael Redclift and Graham Woodgate, 138–149. Cheltenham: Edward Elgar.

2001. *Globalization and Environmental Reform: The Ecological Modernization of the Global Economy*. Cambridge, MA: MIT Press.

Myers, Norman. 1992. "The Anatomy of Environmental Action: The Case of Tropical Deforestation." In *The International Politics of the Environment:*

Actors, Interests, and Institutions, edited by Andrew Hurrell and Benedict Kingsbury, 430–454. Oxford: Clarendon Press.

Nakayama, Mikiyasu. 2000. "The World Bank's Environmental Agenda." In *The Global Environment in the Twenty-First Century: Prospects for International Cooperation*, edited by Pamela S. Chasek, 399–410. New York: United Nations University Press.

Nash, Roderik. 1973. *Wilderness and the American Mind*. New Haven: Yale University Press.

National Research Council (NRC). 2010. *Advancing the Science of Climate Change*. Washington, DC: National Academies Press.

Newell, Peter. 2000. *Climate for Change: Non-state Actors and the Global Politics of the Greenhouse*. Cambridge: Cambridge University Press.

Norgaard, Kari Marie. 2011. *Living in Denial: Climate Change, Emotions, and Everyday Life*. Cambridge, MA: MIT Press.

North, Douglass C. 1990. *Institutions, Institutional Change, and Economic Performance*. Cambridge: Cambridge University Press.

Oberschall, Anthony. 1973. *Social Conflict and Social Movements*. Englewood Cliffs: Prentice-Hall.

O'Connor, J. 1991. "On the Two Contradictions of Capitalism." *Capitalism Nature Socialism* 2 (3): 107–109.

O'Neill, Michael. 1997. *Green Parties and Political Change in Contemporary Europe: New Politics, Old Predicaments*. Aldershot: Ashgate.

Parson, Edward A. 2003. *Protecting the Ozone Layer: Science and Strategy*. Oxford: Oxford University Press.

Paterson, Matthew. 1996. *Global Warming and Global Politics*. New York: Routledge.

Pellow, David Naguib. 2007. *Resisting Global Toxics: Transnational Movements for Environmental Justice*. Cambridge, MA: MIT Press.

Pepper, David. 1984. *The Roots of Modern Environmentalism*. London: Croom Helm.

Pew Center on Global Climate Change. 2006. *Climate Change 101: State Action*. Arlington, VA: Pew Center on Global Climate Change.

Piven, Frances Fox, and Richard A. Cloward. 1977. *Poor People's Movements: Why They Succeed, How They Fail*. New York: Pantheon.

Ponting, Clive. 1991. *A Green History of the World*. London: Sinclair-Stevenson.

Porter, Michael E., and Claas van der Linde. 2000[1995]. "Green and Competitive: Ending the Stalemate." In *The Dynamics of the Eco-efficient Economy: Environmental Regulation and Competitive Advantage*, edited by Emiel F. M. Wubben, 33–56. Northampton, MA: Edward Elgar.

Princen, Thomas. 1994. "NGOs: Creating a Niche in Environmental Diplomacy." In *Environmental NGOs in World Politics: Linking the Local and the Global*, edited by Thomas Princen and Matthias Finger, 29–47. New York: Routledge.

Pring, George W. 2001. "The United States Perspective." In *Kyoto: From Principles to Practice*, edited by Peter D. Cameron and Donald Zillman, 185–224. Norwell, MA: Kluwer Law International.

Probst, Katherine N., and Thomas C. Beierle. 1999. *The Evolution of Hazardous Waste Programs: Lessons from Eight Countries.* Washington, DC: Resources for the Future.

Rabe, Barry G. 2009. "Second-Generation Climate Policies in the States: Proliferation, Diffusion, and Regionalization." In *Changing Climates in North American Politics: Institutions, Policymaking, and Multilevel Governance,* edited by Henrik Selin and Stacy VanDeveer, 67–86. Cambridge, MA: MIT Press.

Ramirez, Francisco O., Yasemin Nuhoglu Soysal, and Suzanne Shanahan. 1997. "The Changing Logic of Political Citizenship: Cross-National Acquisition of Women's Suffrage." *American Sociological Review* 62 (5): 735–745.

Raustiala, Kal, and David G. Victor. 1998. "Conclusion." In *The Implementation and Effectiveness of International Environmental Commitments: Theory and Practice,* edited by David G. Victor, Kal Raustiala, and Eugene B. Skolnikoff, 659–708. Cambridge, MA: MIT Press.

Ringquist, Evan J., and Tatiana Kostadinova. 2005. "Assessing the Effectiveness of International Environmental Agreements: The Case of the 1985 Helsinki Protocol." *American Journal of Political Science* 49 (1): 86–102.

Risse, Thomas, and Kathryn Sikkink. 1999. "The Socialization of International Human Rights Norms into Domestic Practices: Introduction." In *The Power of Human Rights: International Norms and Domestic Change,* edited by Thomas Risse, Stephen C. Ropp, and Kathryn Sikkink, 1–38. Cambridge: Cambridge University Press.

Roberts, J. Timmons, and Peter E. Grimes. 2002. "World System Theory and the Environment: Toward a New Synthesis." In *Sociological Theory and the Environment: Classical Foundations, Contemporary Insights,* edited by Riley E. Dunlap, Frederick H. Buttel, Peter Dickens, and August Gijswijt, 167–196. Lanham: Rowman and Littlefield.

Rootes, Christopher. 2003a. "The Transformation of Environmental Activism: An Introduction." In *Environmental Protest in Western Europe,* edited by Christopher Rootes, 1–19. Oxford: Oxford University Press.

———. 2003b. "Britain." In *Environmental Protest in Western Europe,* edited by Christopher Rootes, 20–58. Oxford: Oxford University Press.

Rosa, Eugene, Andreas Diekmann, Thomas Dietz, and Carlo C. Jaeger. 2010. *Human Footprints on the Global Environment: Threats to Sustainability.* Cambridge, MA: MIT Press.

Rosecrance, Richard N. 1963. *Action and Reaction in World Politics: International Systems in Perspective.* Boston: Little, Brown.

Rowland, Wade. 1973. *The Plot to Save the World.* Toronto: Clarke, Irwin, and Company.

Rowlands, Ian H. 1995. *The Politics of Global Atmospheric Change.* Manchester: Manchester University Press.

Rudel, Thomas K. 1993. *Tropical Deforestation: Small Farmers and Land Clearing in the Ecuadorian Amazon.* New York: Columbia University Press.

———. 2005. *Tropical Forests: Regional Paths of Destruction and Regeneration in the Late 20th Century.* New York: Columbia University Press.

Schaefer-Caniglia, Beth. 2001. "Informal Alliances vs. Institutional Ties: The Effects of Elite Alliances on Environmental TSMO Networks." *Mobilization* 6 (1): 37–54.

Schechter, Michael G. 2005. *United Nations Global Conferences*. London: Routledge.

Schnaiberg, Allen. 1980. *The Environment from Surplus to Scarcity*. New York: Oxford University Press.

Schnaiberg, Allen, and Kenneth Gould. 1994. *Environment and Society: The Enduring Conflict*. New York: St. Martin's Press.

Schnaiberg, Allen, David Pellow, and Adam Weinberg. 2002. "The Treadmill of Production and the Environmental State." In *The Environmental State under Pressure*, edited by Arthur P. J. Mol and Frederick H. Buttel, 15–32. Oxford: Elsevier Science.

Schneiberg, Marc, and Elisabeth Clemens. 2006. "The Typical Tools for the Job: Research Strategies in Institutional Analysis," *Sociological Theory* 3: 195–227.

Schofer, Evan, and Ann Hironaka. 2005. "World Society and Environmental Protection Outcomes." *Social Forces* 84 (1): 25–47.

Schofer, Evan, and Francisco Granados. 2006. "Environmentalism, Globalization, and National Economies: Theories and Evidence, 1980–2000." *Social Forces*, 85, 2:965–991.

Schofer, Evan, Ann Hironaka, David John Frank, and Wesley Longhofer. 2012. "Sociological Institutionalism and World Society." In *The Wiley-Blackwell Companion to Political Sociology*, edited by Edwin Amenta, Kate Nash, and Alan Scott, 57–68. New York: Wiley-Blackwell.

Schofer, Evan, and John W. Meyer. 2005. "The Worldwide Expansion of Higher Education in the Twentieth Century." *American Sociological Review* 70 (6): 898–920.

Schutz, Alfred. 1967. *The Phenomenology of the Social World*. Evanston: Northwestern University Press.

Selin, Henrik, and Stacy D. VanDeveer, eds. 2009, *Changing Climates in North American Politics: Institutions, Policymaking, and Multilevel Governance*. Cambridge, MA: MIT Press.

Sewell, William H. 1980. *Work and Revolution in France: The Language of Labor from the Old Regime to 1848*. Cambridge: Cambridge University Press.

Shandra, John M., Christopher Leckband, and Bruce London. 2009a. "Ecologically Unequal Exchange and Deforestation: A Cross-National Analysis of Forestry Export Flows." *Organization and Environment* 22 (3): 293–310.

Shandra, John M., Christopher Leckband, Laura A. McKinney, and Bruce London. 2009b. "Ecologically Unequal Exchange, World Polity, and Biodiversity Loss: A Cross-National Analysis of Threatened Mammals." *International Journal of Comparative Sociology* 50 (3–4): 285–310.

Shandra, John M., Eran Shor, and Bruce London. 2008. "Debt, Structural Adjustment, and Organic Water Pollution." *Organization and Environment* 21 (1): 38–55.

——— 2009. "World Polity, Unequal Ecological Exchange, and Organic Water Pollution: A Cross-National Analysis of Developing Nations. *Human Ecology Review* 16 (1): 53–63.

Shorette, Kristen. 2012. "Outcomes of Global Environmentalism: Longitudinal and Cross-National Trends in Chemical Fertilizer and Pesticide Use." *Social Forces* 91 (1): 299–325.

Sitarz, Daniel, ed. 1993. *Agenda 21: The Earth Summit Strategy to Save Our Planet*. Boulder: Earthpress.

Sklair, Leslie. 1994. "Global Sociology and Global Environmental Change." In *Social Theory and the Global Environment*, edited by Michael Redclift and Ted Benton, 205–227. New York: Routledge.

Skocpol, Theda. 1992. *Protecting Soldiers and Mothers: The Political Origins of Social Policy in the United States*. Cambridge, MA: Harvard University Press.

Smith, Jackie. 1997. "Building Political Will after UNCED: EarthAction International." In *Transnational Social Movements and Global Politics*, edited by Jackie Smith, Charles Chatfield, and Ron Pagnucco, 175–191. Syracuse: Syracuse University Press.

Smith, S. J., J. van Aardenne, Z. Klimont, R. J. Andres, A. Volke, and S. Delgado Arias. 2011. *Anthropogenic Sulfur Dioxide Emissions, 1850–2005: National and Regional Data Set*, Version 2.86. New York: NASA Socioeconomic Data and Applications Center (SEDAC).

Snow, David, and Robert Benford. 1988. "Ideology, Frame Resonance, and Participant Mobilization." *International Social Movements Research* 1:197–217.

Snow, David, E. B. Rochford, S. Warden, and Robert Benford. 1986. "Frame Alignment Processes, Micromobilization, and Movement Participation." *American Sociological Review* 51:464–481.

Sonnenfeld, D. A. 2000. "Contradictions of Ecological Modernization: Pulp and Paper Manufacturing in South-East Asia." In *Ecological Modernisation around the World: Perspectives and Critical Debates*, edited by A. P. J. Mol and D. A. Sonnenfeld, 235–256. Essex: Frank Cass.

Soroos, Marvin S. 1997. *The Endangered Atmosphere: Preserving a Global Commons*. Columbia: University of South Carolina Press.

Soysal, Yasemin Nuhoglu. 1994. *Limits of Citizenship: Migrants and Postnational Membership*. Chicago: University of Chicago Press.

Spaargaren, Gert, and Arthur P. J. Mol. 1992 "Sociology, Environment, and Modernity: Ecological Modernization as a Theory of Social Change." *Society and Natural Resources* 5:323–344.

Statesman's Yearbook. 1950–2010. London: Palgrave.

Stoel, Thomas B., Jr., Alan S. Miller, and Breck Milroy. 1980. *Fluorocarbon Regulation: An International Comparison*. Lexington, MA: DC Heath and Company.

Strang, David. 2010. *Learning by Example: Imitation and Innovation at a Global Bank*. Princeton: Princeton University Press.

Strang, David, and John W. Meyer. 1993. "Institutional Conditions for Diffusion." *Theory and Society* 22 (4): 487–511.

Suchman, Mark C., and Dana P. Eyre. 1992. "Military Procurement as Rational Myth: Notes on the Social Construction of Weapons Proliferation." *Sociological Forum* 7 (1): 137–161.

Susskind, Lawrence. 1994. *Environmental Diplomacy: Negotiating More Effective Global Agreements*. Oxford: Oxford University Press.

Susskind, Lawrence, and Connie Ozawa 1992. "Negotiating More Effective International Environmental Agreements." In *The International Politics of the Environment: Actors, Interests, and Institutions*, edited by Andrew Hurrell and Benedict Kingsbury, 142–165. Oxford: Clarendon Press.

Swidler, Ann. 1986. "Culture in Action: Symbols and Strategies." *American Sociological Review* 51 (2): 273–286.

——. 2001. *Talk of Love: How Culture Matters*. Chicago: University of Chicago Press.

Szasz, Andrew. 1994. *EcoPopulism: Toxic Waste and the Movement for Environmental Justice*. Minneapolis: University of Minnesota Press.

Taylor, Bron, Heidi Hadsell, Lois Lorentzen, and Rik Scarce. 1993. "Grass-Roots Resistance: The Emergence of Popular Environmental Movements in Less Affluent Countries." In *Environmental Politics in the International Arena: Movements, Parties, Organizations, and Policy*, edited by Sheldon Kamieniecki, 69–90. Albany: State University of New York Press.

Tilly, Charles. 1978. *From Mobilization to Revolution*. Reading: Addison-Wesley.

Tolba, Mostafa K. 1982. *Development without Destruction: Evolving Environmental Perceptions*. Dublin: Tycooly International Publishing.

Tolba, Mostafa K., and Iwona Rummel-Bulska. 1998. *Global Environmental Diplomacy: Negotiating Environmental Agreements for the World, 1973–1992*. Cambridge, MA: MIT Press.

Tolba, Mostafa K., and Osama A. El-Kholy, eds. 1992. *The World Environment, 1972–1992: Two Decades of Challenge*. New York: Chapman and Hall.

Tsutsui, Kiyoteru, and Hwa-Ji Shin. 2008. "Global Norms, Local Activism, and Social Movement Outcomes: Global Human Rights and Resident Koreans in Japan." *Social Problems* 55 (3): 391–418.

Tsutsui, Kiyoteru, and Christine Min Wotipka. 2004. "Global Civil Society and the International Human Rights Movement: Citizen Participation in Human Rights International Nongovernmental Organizations." *Social Forces* 83 (2): 587–620.

Union of International Associations (UIA). 1970–2010. *Yearbook of International Organizations*. Brussels: UIA.

United Nations Conference on the Human Environment (UNCHE). 1972. *Declaration of the United Nations Conference on the Human Environment*. Stockholm: 5–16 June.

United Nations Environment Programme (UNEP). 2005. *Register of International Treaties and Other Agreements in the Field of the Environment*. Nairobi: UNEP.

Vajpeyi, Dhirendra K., ed. 2001. *Deforestation, Environment, and Sustainable Development*. Westport: Praeger.

Van den Akker, J. L. 2000. "Comment: Integrating the Environment in Business Practices." In *The Dynamics of the Eco-efficient Economy: Environmental Regulation and Competitive Advantage*, edited by Emiel F. M. Wubben, 133–138. Northampton, MA: Edward Elgar.

Victor, David G., Kal Raustiala, and Eugene B. Skolnikoff. 1998. "Introduction and Overview." In *The Implementation and Effectiveness of International Environmental Commitments: Theory and Practice*, edited by David G. Victor, Kal Raustiala, and Eugene B. Skolnikoff, 1–46. Cambridge, MA: MIT Press.

Vogel, David. 1995. *Trading Up: Consumer and Environmental Regulation in a Global Economy*. Cambridge, MA: Harvard University Press.

Vogel, David, and Timothy Kessler. 1998. "How Compliance Happens and Doesn't Happen Domestically." In *Engaging Countries: Strengthening Compliance with International Environmental Accords*, edited by Edith Brown Weiss and Harold K. Jacobson, 19–38. Cambridge, MA: MIT Press.

Wallerstein, Immanuel. 1999. "Ecology and Capitalist Costs of Production: No Exit." In *Ecology and the World-System*, edited by Walter L. Goldfrank, David Goodman, and Andrew Szasz, 3–12. Westport: Greenwood Press.

Walsh, Edward J., Rex Warland, and D. Clayton Smith. 1997. *Don't Burn It Here: Grassroots Challenges to Trash Incinerators*. University Park: Pennsylvania State University Press.

Wapner, Paul. 1996. *Environmental Activism and World Civic Politics*. Albany: State University of New York Press.

——— 2000. "The Transnational Politics of Environmental NGOs: Governmental Economic and Social Activism." In *The Global Environment in the Twenty-First Century: Prospects for International Cooperation*, edited by Pamela S. Chasek, 87–108. New York: United Nations University Press.

Ward, Barbara, and René Dubos. 1972. *Only One Earth: The Care and Maintenance of a Small Planet*. New York: W.W. Norton and Company.

Weick, Karl E. 1976. "Educational Organizations as Loosely Coupled Systems." *Administrative Science Quarterly* 21 (1): 1–19.

Weidner, Helmut, and Martin Jänicke, eds. 2002. *Capacity Building in National Environmental Policy*. New York: Springer.

Weiner, Douglas R. 2006. "Environmental Activism in the Soviet Context: A Social Analysis." In *Shades of Green: Environmental Activism around the Globe*, edited by Christof Mauch, Nathan Stoltzfus, and Douglas R. Weiner, 101–134. Lanham: Rowman and Littlefield.

Westney, D. Eleanor. 1987. *Imitation and Innovation: The Transfer of Western Organizational Patterns to Meiji Japan*. Cambridge, MA: Harvard University Press.

World Bank. 2013. *World Development Indicators*. Washington, DC: World Bank.

Wotipka, Christine Min, and Kiyoteru Tsutsui. 2008. "Global Human Rights and State Sovereignty: State Ratification of International Human Rights Treaties, 1965–2001." *Sociological Forum* 23 (4): 724–754.

Wubben, Emiel F. M. 2000. "The Eco-efficient Economy: Threat or Opportunity for Companies?" In *The Dynamics of the Eco-efficient Economy: Environmental Regulation and Competitive Advantage*, edited by Emiel F. M. Wubben, 1–32. Northampton, MA: Edward Elgar.

Yearley, Steven. 1991. *The Green Case: A Sociology of Environmental Issues, Arguments, and Politics*. London: Harper Collins Academic.

York, Richard, Eugene A. Rosa, and Thomas Dietz. 2003. "Footprints on the Earth: The Environmental Consequences of Modernity." *American Sociological Review* 68 (2): 279–300.

Young, Oran R. 1989a. "Science and Social Institutions: Lessons for International Resource Regimes." In *International Resource Management: The Role of*

Science and Politics, edited by Steinar Andresen and Willy Østreng, 7–24. New York: Belhaven Press.

1989b. "The Politics of International Regime Formation: Managing Natural Resources and the Environment." *International Organization* 43:349–375.

Zald, Mayer N., and John McCarthy. 1980. "Social Movement Industries: Competition and Cooperation among Movement Organizations." In *Research in Social Movements, Conflict, and Change*, Vol. 3., edited by Louis Kriesberg, 1–20. Greenwich: JAI Press.

Zelizer, Viviana A. Rotman. 1994. *Pricing the Priceless Child*. Princeton: Princeton University Press.

Index

acid rain, 37, *see also* sulfur dioxide
actors, 78, 82, 87, 152
  definition of, 78
agenda setting, 60–62, 67–70
  national agendas, 61, 69–70
agents, 17, 20–21, 64, 79, 115, 141, 154,
    155, 156–157, 161, *see also* Chapter 4
  definition of, 78
anachronism, 123, 124

Basel Convention on the Transboundary
  Movement of Hazardous Wastes and
  Their Disposal, 100
Bee Swarm, 17, 20, 49, 77, 84, 86, 98, 119,
    123, 140, 143, 151, 155, 159, 160, 168
  definition of, 7–9
Berger and Luckmann, 16, 119, 139
boundaries, institutional, 118, 120
Boyle, Elizabeth, 90

carbon dioxide emissions, 163, 164, 166
chlorofluorocarbons (CFCs), 65,
  see also ozone depletion
climate change, 126–136
  United States, 61, 125, 130, 132–135
conflict between institutions, 21–22,
    120–121, 136
contingency, 22, 116, 117–119, 154
control variables in analyses, 165, 166, 168
Cooley, Charles Horton, 16, 79
cultural meaning, 8, 18, 139, 143,
    157–158, *see also* Chapter 5
  changes in, 115–117

data collection, 61, 69
Declaration of the United Nations
  Conference on the Human
  Environment (UNCHE), 42–46
decoupling, 150, *see also* loose coupling
deforestation, 143–145, 163, 165, 168
desertification, 142–143
development regime, 38–40, 45
diffusion, 51, 89–91, 100, 121–123, 150
discourse, 113, 115
Dobbin, Frank, 88, 90

Earth Day, 32
ecosystemic framework, 25–28, 46,
    47, 144
enforcement theory, 56–57, 62
Enlightenment, 146
Environment Forum, 33

failure of institutions, 22, 140–145,
  see also Chapter 6
Foucault, 115, 159
frames, 107
future research, 123, 154, 155, 156,
    158

Global Climate Coalition, 128, 129, 131
Goffman, Erving, 79
greenhouse gases, 163, *see also* carbon
  dioxide emissions

Hallett and Ventresca, 88
hazardous wastes, 92–102

treaties, environmental, 48, 165, 166,
   168

United Nations, 41–42
United Nations Conference on
   Environment and Development
   (UNCED), 40
United Nations Environment Programme
   (UNEP), 24, 46, 153
United Nations Framework Convention
   for Climate Change (UNFCCC),
   126, 130

values, 138–140
Vienna Protocol, 72

water pollution, 163, 165, 166
workspaces, 19–20, 49–50, 62,
   155
   ozone depletion, 70–73
   Stockholm Conference, 42–46
World Bank, 92
world society theory, 15–18, 112
world systems theory, 12–13
World War II, 146

Made in the USA
Middletown, DE
23 September 2018